CW01022240
Happy 12th Birthday
Fergus
lots of love
from Grandpa Dave
& Granny Rosie
xx

Scottish Fossils

By the same author and also published by Dunedin Academic Press:

Fossils Alive, or New Walks in an Old Field
Excursion Guide to the Geology of East Sutherland and Caithness (Second Edition)

See www.dunedinacademicpress.co.uk for details of these and all our publications

SCOTTISH FOSSILS

Nigel H. Trewin

Department of Geology and Petroleum Geology,
University of Aberdeen
and
National Museums Scotland

EDINBURGH ◆ LONDON

DEDICATION

This book is dedicated to all amateur collectors who continue to advance knowledge of Scottish palaeontology, and particularly to the late Stan Wood for his many important finds and excavations that have contributed so much to science in recent years.

First published in 2013 by
Dunedin Academic Press Ltd

Head Office: Hudson House,
8 Albany Street, Edinburgh EH1 3QB

London Office: The Towers,
54 Vartry Road, London N15 6PU

ISBN 978-1-780460-019-2

British Library Cataloguing in Publication Data
A catalogue record for this book is available from the British Library

Design and pre-press production
by Makar Publishing Production, Edinburgh

Printed in Poland by Hussar Books

Contents

Foreword

This book introduces the reader to a selection from the great variety of fossils found in Scotland. I hope that anybody with an interest in the history of life on Earth will appreciate the illustrations of fossils, and those with deeper interests can delve into the details and references provided. I have chosen the fossils out of thousands of possible candidates to reflect a variety of features. Some are famous fossils, with world-wide scientific significance; others are particularly beautiful as objects, or show exceptional preservation of the organism. My choices also demonstrate the general biological diversity represented by fossils found in Scotland; everything from small oyster shells to giant ammonites; from ancient plants to dinosaur bones, and from primitive armoured fish to superb trilobites. Some of the fossils illustrated are very rare, but I have also included fossils that are common and can be found by amateur collectors visiting the appropriate locality.

Scotland has very diverse geology for such a small country, and this diversity is also reflected in the fossils, which span some 800 million years of Earth history. Several of the fossils have historical associations with places and people, such as the ammonite named after the wife of the nineteenth century geologist Sir Roderick Murchison, and the fossil fish named after Hugh Miller, the Cromarty stone mason. Some fossils, such as the trilobite *Olenellus* from the NW Highlands, have importance in that they provided evidence for the opening and closing of past oceans.

There are some aspects of Scottish palaeontology that are of international renown, and naturally these aspects are given more pages and examples. The Devonian fossil fish of Caithness are some of the best preserved in the world, and the Early Devonian biota of plants and associated arthropods from the Rhynie chert of Aberdeenshire represent the best-preserved early terrestrial ecosystem ever found. Superb fossil plants are found in the Coal Measures of the Midland Valley, and fantastic arthropods looking like giant water scorpions come from rocks at Lesmahagow.

I freely admit that the choices also reflect some of my personal interests, and that another author might choose many different fossils, but I suspect that most would appear on any palaeontologist's list of interesting Scottish fossils. The selection also highlights museum collections, where some of the illustrated fossils are on display, but most are hidden from public view in ranks of drawers in museum stores.

The fossils are arranged according to biological classification, starting with bacteria, algae and plants, and followed by animals in a generally increasing order of complexity from sponges to mammals. I have included details of biological classification, and given localities for most of the fossils illustrated. Remember that many of these localities are protected sites (SSSIs), and permission is needed both for access, and to collect fossils. Many of the localities no longer exist as sites for collecting; for example, abandoned mines and infilled quarries. If you collect fossils please follow the Scottish Fossil Code produced by Scottish Natural Heritage, and collect in a responsible manner. The Scottish Fossil Code includes useful information and advice on collecting, preparing specimens and documenting a collection.

The Code can be viewed at www.snh.gov.uk/docs/B572665.pdf.

Acknowledgements

It would not have been possible to produce this book without the assistance of curators and archivists in the museums that have contributed images, and allowed me to photograph specimens in their care. The collections of National Museums, Scotland under the care of Nick Fraser, and in the Hunterian Museum, University of Glasgow, under the care of Neil Clark, have provided numerous specimens. I thank them for their help and patience with my many enquiries. The following have also rendered excellent service in finding museum specimens and images: Andrew Ross, Stig Walsh, Yves Candela and Maggie Wilson at National Museums Scotland, Edinburgh; Graham Nisbet and Neil Clark at the Hunterian Museum, Glasgow; Bill Dalgarno and Janet Trythall at Elgin Museum; David Lampard at the McManus Art Gallery and Museum, Dundee; Jason Sutcliffe of East Ayrshire Council, Dick Institute, Kilmarnock; Stuart Allison, St Andrews University; Dugald Ross, Isle of Skye Museum; the British Geological Survey and Vicki Hammond at The Royal Society of Edinburgh.

The following individuals have also kindly allowed me to use images from private collections, or helped with providing fossil reconstructions and identifications: Brian Bell, Glasgow University; Dmitry Bogdanov; Yves Candela, NMS; Neil Clark, Hunterian Museum, Glasgow; Jenny Clack, Cambridge University:

Euan Clarkson, Edinburgh University; Mike Coates, University of Chicago; Bob Davidson, Aberdeen University; David Harper, Durham University; Carol Hopkins; Roger Jones; Hans Kerp, Münster University; Laura Säilä, University of Helsinki and John Whicher. Acknowledgements for individual images, and collection locations of specimens, are given following the specimen descriptions section of the book.

Finally, I must mention the numerous collectors who found the fossils, for without them there would be no museum collections. The specimens illustrated in this book have collection dates ranging from pre-1850 (Hugh Miller Collection) right up to the present day. In recent times there have been many exciting new finds in Scottish palaeontology, several discovered through the site research and excavations instigated by the late Stan Wood. Palaeontology is a science where expert amateurs can make a valuable contribution, and several have contributed to this book. Their enthusiasm and effort continues to advance Scottish palaeontology.

Guide to the information given for each fossil

Each fossil in the book carries a simple heading with the fossil name (genus), the general type of fossil (e.g. trilobite, ammonite, plant, fish, reptile) and the Geological Period in which the chosen specimen was living. The following more detailed information is then given for each fossil.

Phylum

The Animal Kingdom is divided into major groups known as Phyla (e.g. Mollusca; the molluscs include clams, oysters, snails, slugs, octopus, squid). The Plant Kingdom is separated into Divisions rather than Phyla.

Class

The first subdivision of a phylum. A Class of the Mollusca is Bivalvia. This comprises the bivalves (e.g. clams, oysters, scallops).

Order, Family

Further subdivisions are made dividing the members of each Class into Orders and Families; most of these subdivisions are omitted in this book.

Genus and Species

This is the binomial name that defines each described species of plant or animal. The common large scallop is called *Pecten maximus*. *Pecten* is the generic name and *maximus* the specific name. Most fossils are known by these latinised names because few fossils have common names. Hence your pet dog is *Canis familiaris* in technical terminology. For many of the fossils in this book specific names have not been used. The names applied to fossils can change as palaeontologists publish taxonomic revisions. I have generally used the names on the labels of museum specimens, some of which may not reflect the latest publications.

Locality

This is the place where the illustrated specimen was found. The fossil may also be found at other localities where rocks of the same age are present. A fossil that lacks a record of the locality where it was collected is generally of little scientific value. If you collect a fossil, always record the locality details.

Age

This is generally expressed as the **Geological Period**, such as Jurassic or Devonian. In some cases the Stage name within the Period is also given.

Stratigraphy

Sequences of sedimentary rocks can be divided on the basis of time intervals in which they were deposited (Chronostratigraphy) or on the basis of lithologies (Lithostratigraphy). Thus the Devonian Period is defined as an interval of time, but the Old Red Sandstone is a lithological division, and whilst it is mainly of Devonian age, in some places it extends down into the Silurian, or up into the Carboniferous. The subdivisions of the rock record are constantly being revised, and there are probably hundreds of names used for rock sequences in Scotland. In descending order the terms used are **Groups, Formations, Members, Units and Beds.** Thus this information records the **Stratigraphic Position** of the fossil in the rock record. The stratigraphic information available on the label of a museum fossil frequently records an 'old' stratigraphic name, and can be difficult to interpret in terms of modern stratigraphic schemes. As far as possible I have attempted to record the **Formation** in which the fossil was found.

Introduction

This book is a celebration of the fossils of Scotland, and aims to provide the reader with an armchair museum gallery of Scottish fossils. Many of the specimens are rare and scientifically important specimens; others are common but typical representatives. Some have stories to tell regarding the part they have played in the history of science in Scotland and the wider world, whilst others are superb and impressive specimens that would grace any museum collection. There are many excellent books available that describe the various groups of fossils, explaining in detail the zoological classification, the morphological details, and the palaeoenvironments in which they lived. This book does not attempt to repeat that information; thus the reader can consult Cleal and Thomas (1999) for plants, Clarkson (1998) and Wyse Jackson (2010) for the invertebrates, and Benton (2009) for vertebrates. References to academic papers are given to enable the reader to find more detail than is appropriate for inclusion in this volume. A geological timescale (Table 1, p. 8) is provided together with major events and environmental changes that have affected Scotland during the span of geological time represented by the fossils in this book.

Limiting the book to Scottish fossils illustrates the fact that an incredible variety of geology is crammed into the small area that is Scotland. Every geological period during which life existed on Earth is represented in Scotland, and there is an extraordinary wealth of sites that are of international importance to palaeontology. In the early nineteenth century when geology emerged as a science, it was considered a worthy occupation for gentlemen, and ladies, to assemble collections of fossils, rocks and minerals as part of a general interest in natural history. These private collections became the nucleus of material that was gradually gathered into museum collections through sales, gifts and legacies. Hugh Miller, famous for his books such as *The Old Red Sandstone* (1841), *Footprints of the Creator* (1849), and *Testimony of the Rocks* (1857) formed a large fossil collection. Started in his home town of Cromarty with Old Red Sandstone fish and also Jurassic ammonites and belemnites from the local area, his collection grew when he moved to Edinburgh. After his death in 1856 his collection went to the Natural History Museum in Edinburgh, which eventually became part of National Museums, Scotland. Miller's collection was bought for a little over £1000, part paid by public subscription. This price can be compared with the £875 that his house sold for in 1864. Thus fossils were greatly valued at the time, and Miller had a fearsome mantrap to deter theft from his collection! Some of his specimens are now back in Cromarty at the Hugh Miller Museum as a loan from NMS. Further details of Miller's fascinating life are discussed by Michael Taylor in his excellent book on this famous character (Taylor 2007).

In the late nineteenth century amateur collectors, many of them landowners, doctors, and church ministers, made valuable collections and became experts in the palaeontology of local areas. Many acquired the knowledge to publish their findings in scientific papers and books. The Great War of 1914–18 seems to have been a turning point; society changed and the enthusiasm for collecting fossils waned. This also coincided with industrialisation and improved transport links. One factor affecting palaeontology was the demise of small local quarries worked by hand, and the creation of larger mechanised operations. The rise of artificial building materials, particularly concrete for building and paving, also forced the closure of many quarries, and fossil supplies diminished.

The years between the wars, including the Great Depression, saw a decline in interest in fossils and collecting. In many local museums the Victorian displays were still in place, and many museums were closed. After the Second World War (1939–45) the emphasis was on the rebuilding of society and industry; food and petrol were rationed and there was general austerity. When I became interested in fossils around 1954 through the chance gift of a book (*British Fossils* by Duncan Forbes) there was virtually no literature at the popular level to help a beginner. I found some information in second-hand books, spending 15 shillings (75 pence) on a copy of Buckland's *Bridgewater Treatise* from Thorpe's second-hand bookshop in Guildford. It was the Natural History Museum in London that sustained my interest, particularly the identification volumes on British Fossils, and the fossil identification service at the museum.

The last 50 years have seen an exponential rise in available literature in the form of books and digital output. The media, particularly through television, have also made an important contribution to public interest in fossils. Today there is a new upsurge of interest in fossils and in collecting. Many localities have hardly been touched for over a century, and there are new finds to be made. Recent discoveries of new material at localities, such as the Carboniferous amphibians of East Kirkton, and arthropods from the Devonian Rhynie chert, are good illustrations of recent advances. It seems we are in a new age of palaeontological progress, with enthusiastic amateurs making an increasing contribution.

Scientific societies also formed collections, of which a notable survivor is the fine collection of the Elgin and Morayshire Scientific Association. The present Elgin Museum was opened in 1843 to house the Association collection, and it is still operated by the Moray Society. Generally the collections formed by such societies became the basis for larger museum collections such as that of National Museums, Scotland.

The older Scottish universities also have a strong tradition in palaeontology; hence at at Aberdeen the basis of the fossil collection can be traced to the efforts of James Nicol and H.A. Nicholson, both professors of Natural History in the nineteenth century. Palaeontology was a 'state of the art' occupation in that fossils were the only way to determine the relative ages of rocks. In England, William Smith (1816) demonstrated that different rock strata could be recognised on the basis of fossils, and from that time, right up to the present, much effort has gone into stratigraphic palaeontology, devising increasingly detailed palaeontological methods of correlation and biostratigraphic dating using fossils. At present, micropalaeontology, using foraminifera and plant spores and pollen, is used extensively for biostratigraphic correlation. This relies on the mechanical or chemical extraction of microfossils from rocks, and is in constant use to monitor drilling progress of oil wells. The great advantage of micropalaeontology is that a large quantity of microfossil material can be extracted from a small volume of rock.

When in 1854 the Geological Survey was extended to Scotland, it was an essential part of mapping that fossils were collected and identified to determine the stratigraphic age of the rocks being mapped. The Survey employed 'fossil collectors' who were sent to make representative collections at locations that had been identified as fossiliferous by geological surveyors. Thus, after William Mackie discovered the Rhynie chert in 1912 (Mackie 1913), the Survey sent Mr Tait to dig trenches and collect material. Tait's material from Rhynie is still in the Geological Survey collections in Edinburgh.

Variety and range of fossils found in Scotland

Within this book many fossil localities, or areas, have several entries, and represent particularly rich and well-preserved faunas and floras. Some are so prolific that a whole book could be devoted to the fossils of a single locality. The following brief notes introduce the reader to some of these localities; further details are given in the text accompanying the individual entries in the book. These localities are recorded in ascending stratigraphic order.

Cambrian of the NW Highlands

The connection between the NW Highlands and the Appalachians of North America was made by Lapworth (1888) through the recognition of the trilobite *Olenellus*, characteristic of Lower Cambrian rocks on both sides of the Atlantic. The classic work of the Geological Survey culminated in the memoir on the structure of the NW Highlands (Peach *et al.* 1907).

Ordovician/Silurian of the Girvan area

The Girvan area has produced fine collections of invertebrates from the Silurian and Ordovician. In particular the Ordovician 'Starfish Bed' of South Threave contains many forms, particularly of echinodermata (starfish, crinoids, and other primitive types), and also trilobites and brachiopods. Large collections made by Mrs Elizabeth Gray (1831–1924) and her family are now in the Hunterian Museum, University of Glasgow, National Museums Scotland in Edinburgh, and the Natural History Museum, London. Nicholson and Etheridge (1878–80) produced a fine monograph on Girvan fossils based on the Gray Collection, and Charles Lapworth (1882) published a classic account of the Girvan succession. A fine selection of fossils from the 'Starfish Bed' is on display in the Hunterian Museum. Harper (1982a, b) discusses the starfish beds, and the stratigraphy of the area, whilst a general account of the area is included in Bluck (2002).

Ordovician/Silurian of Dob's Linn

The graptolitic shales of Dob's Linn near Moffat were studied in detail by Charles Lapworth (1878) and his graptolite zones provided the basis for the unravelling of the stratigraphy of the Southern Uplands in the Geological Survey memoir on *The Silurian Rocks of Britain* by Peach and Horne (1899). Interpretations of the structure of the Southern Uplands have changed with time and have been summarised in *The Geology of Scotland* (Oliver *et al.* 2002). An excursion guide to the locality is provided by Clarkson and Taylor (1992).

Silurian inliers, Pentlands and Lesmahagow

In the Silurian inliers of the Pentland Hills, Hagshaw Hills and Lesmahagow, evidence for the closing of the Iapetus Ocean can be seen. As the Iapetus Ocean, which had separated the faunas of Scotland from those of the rest of the UK gradually closed, the faunas from different sides of the ocean became mixed as they became geographically closer. During ocean closure some areas of sea became cut off from the open oceans, and in those areas the marine faunas died out as the water became less saline. By latest Silurian times these areas had become terrestrial with rivers depositing sandstones and conglomerates of the Old Red Sandstone.

The Lesmahagow and Hagshaw Hills inliers are famed for rich faunas of primitive agnathan (jawless) fish such as *Jamoytius, Birkenia, Thelodus* and *Ateleaspis*. There is also a celebrated arthropod fauna of eurypterids from Lesmahagow, including the giant *Slimonia*, together with *Erretopterus* and *Hughmilleria* and the pod-shrimp *Ceratiocaris*. *Slimonia* is named after Dr Robert Slimon who made a collection of the arthropods after being shown them by a local farmhand. They were exhibited at the British Association meeting of 1855, arousing the interest of Murchison, who went to Lemahagow collecting with Slimon.

The Pentland Hills are famed largely for some excellent marine Silurian faunas. Being close to Edinburgh, this area has received a lot of attention for its brachiopod communities, starfish beds, trilobites and much more. The trilobites have been championed by Euan Clarkson for many years, and his *Pentland Odyssey* (Clarkson 2000) gives a fine summary of Pentlands palaeontology. There is an excellent guide to the fossils of the Pentland Hills by Clarkson *et al.* (2008) from which the collector can identify fossil finds.

Early Devonian, Forfar and Dundee

The fossil localities of the Lower Old Red Sandstone in the area from Forfar to Dundee yield world-renowned acanthodian and cephalaspid fish faunas and also early land plants and arthropods. The acanthodian fauna is particularly impressive in the variety of genera and the preservation of articulated specimens. They were collected by James Powrie (1814–1895) when quarries were active, but material can still be found in old quarry tips, providing evidence for an environment of rivers and lakes, disrupted at times by volcanic activity (Trewin and Davidson 1996). If you find one fossil fish in a full day searching you are lucky.

Early Devonian, the Rhynie chert

The Rhynie chert of Aberdeenshire was discovered by William Mackie in 1912 (Mackie 1913) The plants were initially studied by Kidston and Lang (1916–1921) in a classic series of papers. Over the last 20 years many new discoveries of animals and plants have been made at Rhynie through drilling and trenching at the site. Many of the results of these investigations are published in Trewin and Rice (eds 2004). The chert formed from the silica-rich sinters deposited from hot springs and geysers in a terrestrial environment. Plants, fungi and arthropods are preserved in fantastic three-dimensional cellular detail. All the fossils have to be studied in microscope sections. This deposit contains the earliest known records on Earth of animals as diverse as harvestman spiders and nematode worms. Rhynie is a world-renowned locality that provides a snapshot in time of the biota of an early terrestrial and freshwater ecosystem. A website (Trewin *et al.* 2001) is devoted to the geology and fossils of the Rhynie chert.

Mid-Devonian fish of the Orcadian Basin

The fossil fish of the Middle Old Red Sandstone of Cromarty, Caithness and Orkney are some of the best-preserved Middle Devonian fish known. Initially brought to notice by Hugh Miller in *The Old Red Sandstone* (1841), they rapidly attracted the attention of the Swiss geologist Louis Agassiz, who described many finds in his memoir *Poissons Fossiles du Vieux Grès Rouge* (1844). Most of Hugh Miller's finds came from Cromarty and the Black Isle, but recently new fish have been found in the flagstone cliffs and quarries of Caithness and Orkney. Achanarras Quarry in Caithness provides the greatest variety of well-preserved fish (Trewin 1986). The fish died in mass mortalities and the carcasses eventually sank into the deep anoxic bottom waters of the lake where they were preserved in laminated mud. Scavengers were absent at the bottom of the lake, and currents were very weak, thus the carcasses were gently covered with mud and fine details of delicate fins were preserved. The Achanarras fish bed lamination represents annual climatic periods, and by counting laminations it can be estimated that the fish bed at Achanarras was deposited over a period of about 4000 years. For an excursion guide to the Old Red Sandstone of Caithness, including Achanarras Quarry, *see* Trewin (2009).

Late Devonian of Dura Den, Fife

The Upper Old Red Sandstone fish of Dura Den in Fife were first recorded in 1830, but it was a find of a complete *Holoptychius* shown to the Revd Anderson by a mason in 1836 that really started the excitement. Agassiz described and named the specimen *Holoptychius andersoni*. The quarrymen rapidly realised that money could be made from fossil fish, and that which had previously been destroyed was now saved for collectors. Anderson clearly loved the resulting fame, and his monograph on Dura Den (Anderson 1859) is a Victorian masterpiece of social name-dropping. Charles Lyell and Hugh Miller visited, but his greatest pride seems to have been a visit in 1858 with Sir Roderick Murchison and Lord and Lady Kinnaird and a 'distinguished party from Rossie Priory'.

The fish were crowded on bedding surfaces and represent mass mortalities caused by fish being stranded in pools that dried out.

The general sedimentology of the Upper ORS of the area has been summarised by Trewin and Thirlwall (2002); the sandstones of Dura Den were deposited by both rivers and aeolian processes. Specimens can be seen in many museums, and clearly a lot was excavated and saved. Sadly, the quarries are long since closed and fossils cannot be found at present.

Early Carboniferous of East Kirkton

The East Kirkton Limestone with its interlamination of limestone and chert had been a subject of discussion as long ago as 1825 when it featured in debates between Neptunists and Plutonists. Cadell (1925) recognised that the rocks resembled hot-spring deposits, and overlying tuffs indicate local volcanicity. Some fossils had been found in the early days including *Hibbertopterus,* the first British eurypterid to be described. However, the quarry attracted little attention until 1984 when Stan Wood discovered tetrapod fossils at the site, particularly *Westlothiana,* thought at the time to be the oldest known reptile. Stan Wood's finds prompted a detailed study on all aspects of the geology and fossils of East Kirkton Quarry. The volume edited by Rolfe *et al.* (1994) contains papers on the depositional environment and descriptions of the fauna and flora. The environment was a small lake in a volcanic area influenced by hydrothermal springs that gave the waters an unusual chemistry and resulted in the lack of a normal freshwater fauna. For periods it was not inhabited by aquatic animals, but terrestrial plants and animals were washed or fell into the lake and were preserved. Later in the history of this lake the water became more normal and fish appeared in the lake, indicating that it had become connected to a larger body of water. The East Kirkton Project involving cored drilling and careful bed-by-bed excavation demonstrates how much new information can be gathered when time, finance and expertise are devoted to a project. It also demonstrates how much more there is to find at localities that have received little attention in the past.

Carboniferous shrimps, sharks and a conodont animal

The Carboniferous of the Midland Valley contains shaly sequences in which some beds are so rich in organic matter that they are termed oil shales. The oil shales were the raw material of the oil-shale industry in the Lothians, of which all that remains are red bings of burnt shale. The oil shales and associated limestones and sandstones were deposited in lacustrine to lagoonal settings with variable connection to open marine conditions. Some beds contain abundant shrimps, and rarer fish. One example is the Granton Shrimp Bed that produced *Cladognathus,* the first 'conodont animal'. Another is the fauna of the 'Bearsden Shark Dig' conducted by Stan Wood. This excavation produced a wonderful array of fish, including the strange shark *Akmonistion* as well as the predatory shrimp *Palaemysis.*

Carboniferous flora

Throughout the Carboniferous there is a rich record of the floras, particularly those of the coal forests. The forests of giant horsetails and seed ferns would have been every bit as impressive as modern equatorial rain forest. Robert Kidston (1852–1924) made detailed studies of the Carboniferous floras of Britain, culminating in his detailed memoirs of the Geological Survey published in 1923–1925. His collection of 7500 plant specimens, along with detailed notebooks, was bequeathed to the Geological Survey. He also assisted in the excavation of the 'Fossil Grove' in Victoria Park, Glasgow in 1887–8. Thomson and Wilkinson (2009) provide a biography of Kidston, and for an excellent identification guide to Coal Measures plants *see* Cleal and Thomas (1994).

The Elgin Reptiles

The Permo-Triassic 'New Red Sandstone' of Elgin, Lossiemouth, Burghead and Hopeman is exposed in quarries and coastal cliffs. In the first half of the nineteenth century it was thought that these rocks belonged to the Devonian 'Old Red Sandstone', since Devonian fish had been found at localites such as Tynet Burn near Fochabers, Scaat Craig near Fogwatt and in quarries close to Elgin itself. Local collectors such as Patrick Duff (Town Clerk), John Martin (Teacher) and George Gordon (Minister of the Church) made collections of fossil fish, and formed the Elgin and Morayshire Scientific Association in 1836. The society opened Elgin Museum in 1843 to house the fossils and other collections. The first reptile to be found, *Stagonolepis,* was initially identified as a fish by Agassiz (1844), but the discovery in 1851 of *Leptopleuron,* clearly a small lizard-like reptile, caused great controversy. How could there be reptiles in the Old Red Sandstone?

Following the publicity the find generated, more reptile fossils were found in the quarries. Cutties Hillock Quarry in Quarry Wood, Elgin even had Devonian fish in the base of the quarry and reptiles higher in the same quarry in very similar sandstone.

Eventually a solution was found. In the Elgin area the Old and New red sandstones look very similar, and were deposited in similar environments. An unconformity between the Devonian and Permian sandstone was recognised in Quarry Wood, and by 1859 most experts accepted that there were no reptiles in the Old Red Sandstone.

Permian reptiles, *Elginia, Gordonia* and *Geikia,* occur in the Hopeman Sandstone of Cutties Hillock and Clashach Quarry,

Hopeman. The Triassic fauna of at least eight reptiles includes *Stagonolepis* and *Leptopleuron* and is found in the Lossiemouth Sandstone of quarries at Lossiemouth, Spynie and Findrassie. A lot of reptile specimens were collected in the late nineteenth century when the quarries were working, but finds decreased as many quarries closed in the early twentieth century. The exception is Clashach Quarry at Hopeman, where the dune-bedded sandstones have produced numerous reptile trackways, and a single dicynodont skull. Benton and Walker (1985) discuss the Elgin reptiles in terms of palaeoecology, and McKeever (1994) discusses the reptile trackways of the Permian.

Jurassic Scotland

Jurassic strata occur on both the west and east coasts of Scotland and extend under the North Sea, where they form the main source and reservoir rocks of offshore oilfields. The succession on the Isle of Skye has a variety of marine to freshwater faunas. The marine Jurassic is rich in ammonites, belemnites and bivalves, while the non-marine strata have rare bones and footprints of dinosaurs. In the east coast Jurassic of the Golspie to Helmsdale section shallow to deep water marine environments are represented, and there is evidence that the Helmsdale Fault was active during the Late Jurassic, producing an underwater fault scarp. Earthquakes resulted in deposition of boulder beds that contain faunas swept from shallow water areas over the fault scarp into deep water. It is likely that tsunamis due to earthquake activity were responsible. Hudson and Trewin (2002) provide a summary of the Jurassic of Scotland and further references.

Palaeogene forests and volcanoes

Extensive volcanicity on the west coast of Scotland took place in the Palaeogene with major volcanoes in Ardnamurchan, Mull and Skye. The lava fields and eroded volcanoes are known as the Tertiary Volcanic District. The volcanic activity was connected with the opening of the North Atlantic Ocean, which has widened to its present width in the last 65 million years. In between periods of volcanic eruption forests flourished and are now preserved as leaf beds in sediments between lava flows on Skye and in Mull. The most famous locality is the Ardtun Leaf Bed on Mull, which has been known since 1851, and later in the nineteenth century a quarry was created for the purpose of collecting fossil plants. Boulter and Kvaček (1989) provide a modern account of the plants, concluding that the area was covered with a broad-leaved deciduous forest with only a few conifers. Also on Mull is 'Macculloch's tree', a tree trunk preserved encased in the lava flow that killed the tree, and was first reported by Macculloch in his classic work on the geology of the Western Isles (Macculloch 1819).

Use of fossils in geology and biology

The main geological value of fossils is that they reveal the detailed changes of life on Earth through geological time. As knowledge increases we can subdivide Earth history into smaller and smaller periods. Geological periods are divided into biozones, characterised by different fossils, of which one is chosen as the 'zone fossil'. The requirements of a zone fossil are that it should be widespread, have a short geological range and be easy to recognise. Thus free-swimming marine animals such as ammonites are used as zone fossils in the Jurassic and Cretaceous periods. They evolved so rapidly that changes in faunas can be recognised for periods of much less than a million years. Charles Lapworth (1878) recognised the value of graptolites for biozonation of the Ordovician and Silurian, and made a detailed study of the succession at Dob's Linn near Moffat. This zonation was used by others to interpret the structure of the very thick succession of slates and greywackes of the Southern Uplands. Study of the graptolites enabled the succession to be subdivided, and folding and faulting interpreted.

Early fossil zonations were based on macrofossils that could be identified in the field, but microfossils have been increasingly used because they can be extracted from small samples of rock. Thus foraminifera and plant spores are commonly used to produce biozone schemes, and these are used in industry where material from boreholes is available.

Fossils are essential for the interpretation of ancient sedimentary environments when used in conjunction with sedimentology. Some organisms, such as ammonites and echinoderms, are characteristic of marine conditions; others such as trees and elephants are clearly land-based. The land-based organisms can be washed into the sea after death and may be found preserved near ancient coasts, but it is rare for marine animals to be preserved on land. Fossils also give information on climate in terms of cold- or warm-water faunas. Thus at Hopeman a combination of large-scale dune bedding in sandstones associated with reptile footprints and a reptile skull of Permian age indicates terrestrial dunes. However, sandstones also with cross bedding in the Jurassic at Brora contain scallop shells, ammonites and pieces of fossil wood, indicating a nearshore sandy marine environment.

The history of life on Earth is a worthwhile study in its own right, enabling us to investigate evolutionary sequences in many rapidly evolving groups, and also to study the palaeoecology of ancient fossil communities. In recent years many of the so-called gaps in the fossil record have been filled, such as in the transition from fish to amphibian, where at least six new intermediate forms have been described.

Fossil collecting today

Fossil collecting in Scotland has seen a strong upsurge of interest in the past 30 years, due in a great part to amateur collectors who have sought out old localities and found new material. An abandoned quarry rapidly becomes unproductive for collectors, and spoil tips become overgrown. The Victorian collectors seldom recorded details of the rocks that would help interpret the ancient environment; thus modern techniques can yield valuable new information. Many of the classic collections were assembled by gentlemen buying fossils from quarrymen, and this gives a biased sample of the fauna. Of course, many of the classic localities of the nineteenth century are now lost, largely on account of infilling quarries with waste materials.

An example is the famous Old Red Sandstone fish locality of Achanarras Quarry in Caithness. New species of fish and an arthropod have been discovered by collectors at the site, and the specimens described in scientific papers (Anderson *et al.* 2000; Newman and Trewin 2001, 2008). Scottish Natural Heritage owns the quarry and allows fossil collecting in a responsible manner (according to the Scottish Fossil Code). The spoil tips at the quarry are periodically turned over to maintain the quarry in a fit state for collecting to take place, and a display illustrating the fossil fish, and explaining the geology of the site has been erected in a shelter at the quarry.

Commercial considerations

Like it or not, fossils do have a commercial value. Value may depend on scientific interest, rarity, perfection of the specimen, or display potential. Thus commercial collectors may excavate a site to find and prepare material to sell to museums or private collectors.

The ideal situation might be a commercial excavation that is conducted in such a way that scientific information is recorded in detail, and finds new to science are retained in a museum or other recognised collection where they are available for academic study. Surplus duplicate specimens might then be released for sale.

The worst scenario is the smash and grab raid where a locality is damaged, material is illegally excavated, no records are kept, and all material is sold with vague locality data over the internet or in a rock shop. An example of this activity took place in 2011 at Bearreraig Bay on Skye.

Thankfully, most operations conform more to the former than the latter example given above, and the result has been spectacular for Scottish palaeontology over the last 30 years. In this book you will see reference to the late Stan Wood and the superb finds he made at localities such as Bearsden and East Kirkton. At a time when curators in museums are a diminishing number, and have little time to go into the field and collect specimens, there is a distinct role for the expert commercial collector in the advancement of palaeontology.

There is also a commercial aspect to the expert preparation of specimens. It is good to find a fossil, but the preparation of the specimen may require expertise beyond the ability of the finder. Thus there are preparators who make a career out of fossil preparation for both museums and private individuals. Expert preparation gives fossils 'added value', both scientific and esthetic. This then leads to the manufacture of replicas of fossils, and also making model reconstructions of extinct animals, and hence to computer animation, bringing the past back to life.

Fossil displays in Scotland

Several Scottish institutions have holdings of many thousands of fossils, but only a tiny proportion can be seen in public displays, or even illustrated on museum websites. Hence one purpose of this book is to provide a virtual museum gallery. The following museums have displays of fossils – apologies to any I may have missed.

The Royal Museum of Scotland on Chambers Street, Edinburgh has a series of displays and dioramas in the 'Beginnings' gallery illustrating fossils and environments through geological time, and displaying many excellent specimens. The recent new displays opened in 2011 have taken over the space previously occupied by the old fossil and mineral galleries.

Hunterian Museum, University of Glasgow has excellent collections and displays on a variety of topics, and with plenty to see. Particularly impressive are specimens from the Ordovician 'Starfish bed' in the Girvan area. Other displays cover vertebrate evolution, dinosaurs, and a fine selection of Scottish fossils of all ages. Famous collectors and collections are also featured. A visit is highly recommended to see what are probably the best fossil displays in Scotland, for the variety of specimens on display.

The British Geological Survey holds large fossil collections that are mainly used for research purposes. Web searches can be made for material through the BGS website. A small amount of material is on display in Murchison House, Edinburgh, but is not open to the general public without prior arrangement.

Elgin Museum is a private museum run by the Moray Society. The museum has excellent displays relating to the 'Elgin Reptiles' of the Permo-Triassic, with skeletal material, models and trackways. There are also some very scarce examples of Old Red Sandstone fish on display. The museum provides essential viewing for the famous vertebrate fossil fauna of the area.

The Hugh Miller Museum at Cromarty is a National Trust for Scotland property and has displays on one floor related to Hugh Miller's geological work. Many specimens are on loan from the Hugh Miller Collection held by National Museums, Scotland. The emphasis is on Old Red Sandstone fossil fish, and Jurassic fossils from the local area. There is also a child-friendly table where fossils can be handled. Visitors can also learn about other aspects of Hugh Miller's life, and visit his birthplace cottage next door to the museum.

The Orcadian Stone Company in Golspie has a museum with geological displays of rocks, minerals and fossils mainly from northern Scotland, but also including interesting specimens from world-wide locations.

The Dunrobin Castle Museum is also in Golspie, and in the grounds of Dunrobin Castle. This is a charming example of a Victorian museum that has changed little since it was created. Beyond the stuffed animal trophies, there are some fossils from the local area, many collected in the nineteenth century, but lacking modern labels. Interesting as a historical museum.

The Staffin Museum on the Isle of Skye is the place to go to see some of the dinosaur bones and tracks that have been found in Skye, particularly through the efforts of Dugald Ross.

On Orkney *The Fossil and Heritage Centre* at Burray is privately run, and has on display fine specimens of Old Red Sandstone fish fossils from the Orkney Islands, and other excellent specimens from further afield. New fossil galleries have been recently opened, and are the best fossil displays available in Orkney. There are also good fossils on display in *Stromness Museum.*

Many *town museums* the length and breadth of Scotland have fossil collections, and a selection may be on display, but if all you want to see is fossils it is best to enquire about current exhibits in advance of a visit. A larger selection of such museums is listed in the Scottish Fossil Code, available from Scottish Natural Heritage.

AGE	PERIOD	FOSSILS AND DEPOSITS	LATITUDE	GEOLOGICAL EVENTS
0	QUATERNARY	ICE AGE MAMMALS	57°N	ICE AGES, FORMATION OF GLACIAL TOPOGRAPHY
1.8	NEOGENE	BUCHAN FLINT GRAVELS - DERIVED FOSSILS		EROSION, DEEP WEATHERING, DEPOSITION IN NORTH SEA
24	PALAEOGENE	FLORAS OF MULL AND SKYE WITHIN LAVA PILE		UPLIFT AND EROSION ON LAND RAPID SUBSIDENCE AND SEDIMENTATION, N. SEA BASIN VOLCANIC PROVINCE IN WESTERN SCOTLAND NORTH ATLANTIC OCEAN OPENS
65	CRETACEOUS	CHALK DEPOSITION - SOURCE OF BUCHAN GRAVELS MARINE FAUNA		SCOTLAND BRIEFLY DROWNED BY SEA EROSION OVER MOST OF SCOTLAND DEPOSITION IN OFFSHORE BASINS - MORAY FIRTH
142	JURASSIC	MARINE SHALES, KIMMERIDGIAN, SKYE AND HELMSDALE DINOSAUR TRACKS, BATHONIAN, SKYE MARINE SANDSTONES, AMMONITES, BAJOCIAN, SKYE MARINE LIAS, SKYE, GOLSPIE	40°N	SCOTLAND AN ISLAND, RIFT BASINS IN NORTH SEA AND WEST OF SCOTLAND. SEEN IN HELMSDALE AND SKYE TEMPERATE CLIMATE MAJOR MARINE TRANSGRESSION
205	TRIASSIC	ELGIN REPTILES, LOSSIEMOUTH SANDSTONE	30°N	SCOTLAND IN NORTHERN ARID BELT
248	PERMIAN	ELGIN REPTILES AND TRACKS, HOPEMAN SANDSTONE	20°N	SALT BASINS IN NORTH SEA ERODING LAND AREA WITH SAND DUNES, ELGIN AREA
290	CARBONIFEROUS	FOREST FLORAS THROUGHOUT CARBONIFEROUS BEARSDEN SHARKS, EAST KIRKTON AMPHIBIANS GRANTON SHRIMP BED, CONODONT ANIMAL	10°N	EQUATORIAL FORESTS, RAPID SEA LEVEL CHANGES DUE TO S. HEMISPHERE GLACIATIONS VOLCANIC ACTIVITY IN MIDLAND VALLEY MARINE TRANSGRESSION IN MIDLAND VALLEY
354	DEVONIAN	DURA DEN FISH, UPPER ORS, FIFE ACHANARRAS FISH BED, M.ORS CAITHNESS - ORKNEY FISH AND ARTHROPODS, Lr. ORS, FORFAR - DUNDEE RHYNIE CHERT BIOTA, Lr. ORS, ABERDEENSHIRE	0°	OLD RED SANDSTONE (ORS) DEPOSITED IN CONTINENTAL BASINS WIDESPREAD VOLCANICITY EROSION OF CALEDONIAN MOUNTAINS
418	SILURIAN	EURYPTERIDS AND FISH, LESMAHAGOW MARINE FAUNAS, PENTLAND HILLS	20°S	CLOSURE OF IAPETUS OCEAN AND UPLIFT OF THE CALEDONIAN MOUNTAINS. ISOLATED BASINS IN MIDLAND VALLEY. DEFORMATION OF S. UPLANDS
443	ORDOVICIAN	MARINE FAUNAS, STARFISH BED, GIRVAN AREA DOB'S LINN GRAPTOLITE FAUNAS (ORD-SIL)	30°S	IAPETUS OCEAN CLOSING. MAJOR STRUCTURAL EVENTS SCOTLAND ON LAURENTIA (10°S). ENGLAND ON AVALONIA (65°S). SEPARATED BY IAPETUS OCEAN
495	CAMBRIAN	DURNESS LIMESTONE, NW HIGHLANDS OLENELLUS FAUNA, NW HIGHLANDS	10°S	SCOTLAND PART OF NORTH AMERICA MARINE TRANSGRESSION IN NW HIGHLANDS
545	PRECAMBRIAN	STROMATOLITES c.700Ma		BASEMENT OF LEWISIAN GNEISS (c.3000 Ma), DEFORMED SEVERAL TIMES. TORRIDONIAN SANDSTONE (1200-900 Ma) METAMORPHIC ROCKS OF MOINE AND DALRADIAN

Table 1 Stratigraphic table including geological periods, major fossil-bearing deposits and geological events in Scotland. Part modified from Trewin and Rollin (2002).

BACTERIA

Bacteria formed the oldest fossils on Earth that can be easily recognised in rock outcrops and hand specimens. These are stromatolites that occur in rocks 3.5 billion years old in Western Australia, and they can also be seen in the Precambrian in Scotland on Islay. A variety of bacteria can form stromatolites; some live in hot springs and deposit stromatolites of silica, but most stromatolites are formed in shallow water by cyanobacteria that deposit calcium carbonate. Stromatolites grow in a variety of environments at the present time, the most famous being those in the salty waters of Shark Bay in Western Australia. Examples included here are from the Devonian of Orkney and the Carboniferous of Fife.

1

Stromatolite / Bacteria / Devonian

Phylum	**Cyanobacteria**
Class	
Gen et sp	**No formal name**
Locality	**Stromness, Orkney**
Age	**Middle Devonian**
Stratigraphy	**Lower Stromness Flagstones, Middle Old Red Sandstone**

This example of stromatolites from the Middle Old Red Sandstone of Orkney originally grew on the sediment surface in shallow water near the shore of a lake. Stromatolites require sunlight to grow, so the water would have been clear. There are many laminae within the stromatolite structure, indicating that it probably grew over a period of some years, thus there cannot have been much sediment deposited during this time. The repeated domed growth shapes are typical of stromatolites, and in section can resemble teeth, hence the name 'horse-tooth stone' given to them in Orkney.

1 Stromatolites from the Old Red Sandstone of Orkney.

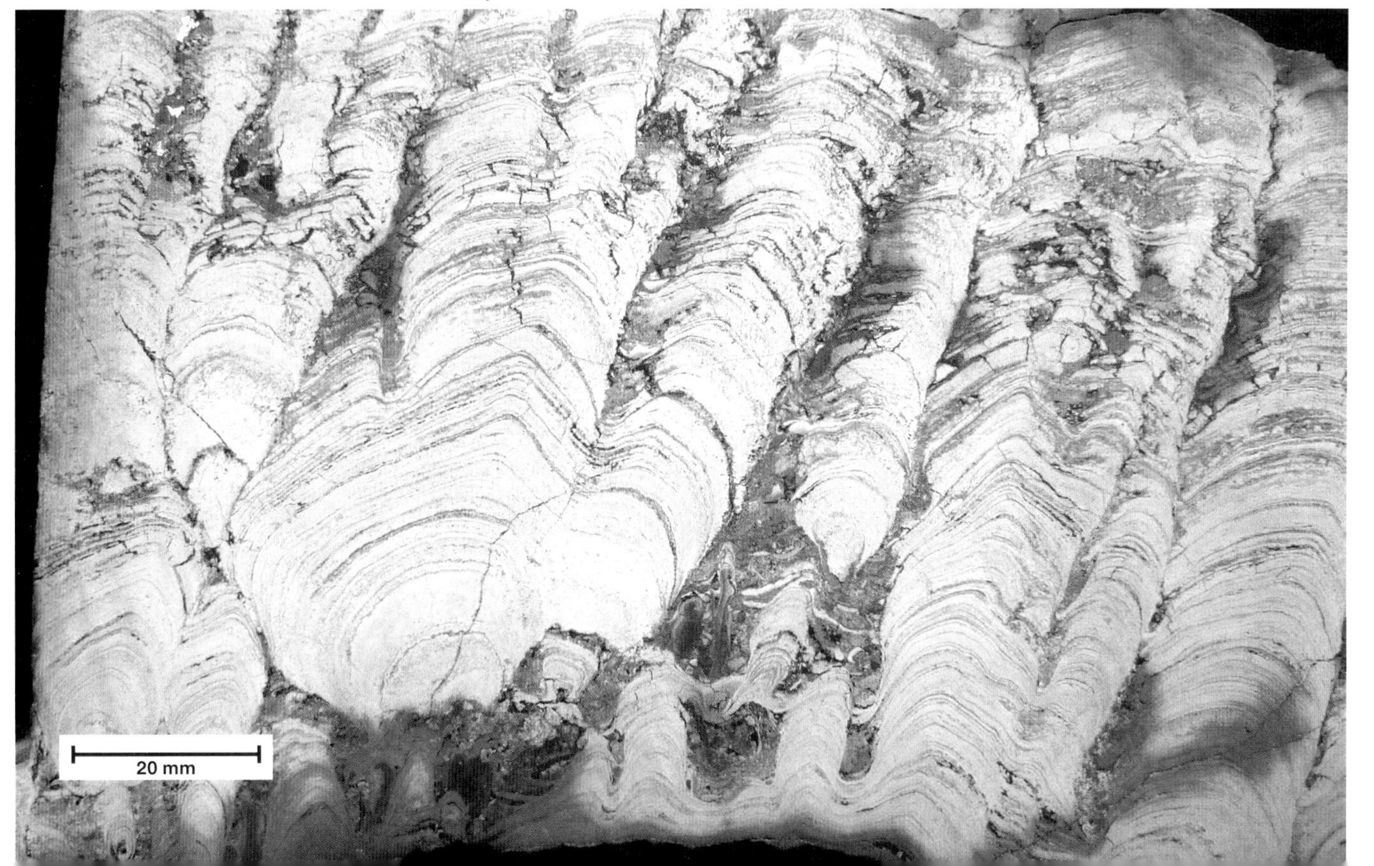

2

Oncolites / Bacteria / Carboniferous

Phylum	**Cyanobacteria**
Class	
Gen et sp	**No formal name**
Locality	**Kingsbarns, Fife**
Age	**Early Carboniferous**
Stratigraphy	**Randerston Limestones, Anstruther Formation, Strathclyde Group**

On the coast to the south of Kingsbarns in Fife a syncline is exposed on the wave-cut platform, and can be viewed at low tide. The rocks are sandstones, shales and thin limestones, and feature in the geological excursion guide to Fife by MacGregor (1973) with further details of the section in the East Fife Memoir (Forsyth and Chisholm 1977). There are 11 limestones in the sequence and most have some fossils, including bivalves and gastropods, and a few of the limestones contain brachiopods. The fauna is of low diversity, and the limestones were not deposited in open marine conditions. The water was probably of variable salinity and the environment one of a delta top with lagoons periodically connected to the sea. Sandstones between the limestones were deposited in river channels on the delta top.

Limestone 9 in the sequence is remarkable in containing red hematite-stained stromatolites with a 'cabbage leaf' form in cross-section. These were fixed during growth, as can be seen by the upward growth of the stromatolite layers in small connected domes. Other structures are round balls of the same material as the stromatolites, and these are called oncolites. The oncolites were not attached and could be rolled about, and thus grew in a spherical form.

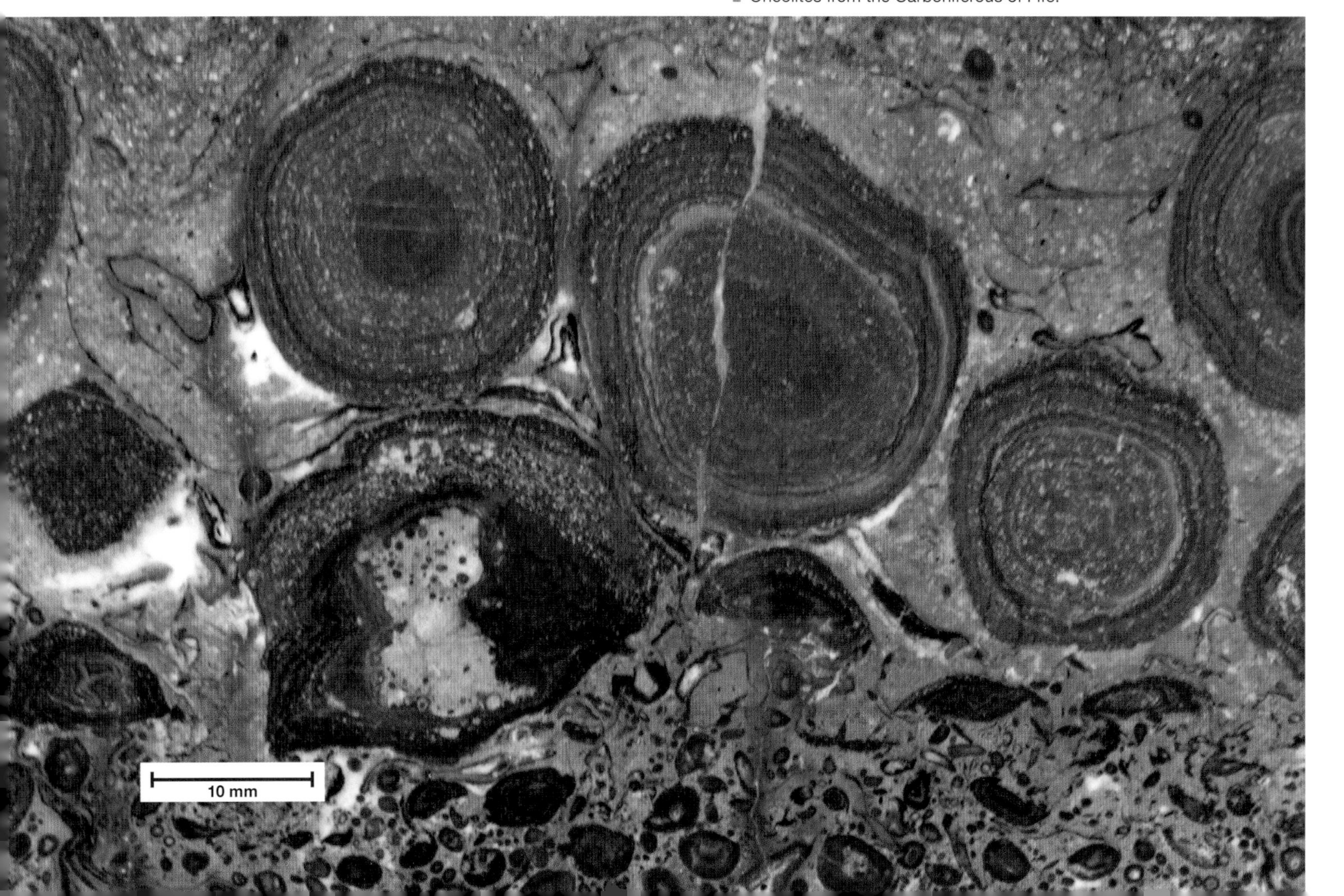

2 Oncolites from the Carboniferous of Fife.

ALGAE

The algae form a highly varied group of organisms, from microscopic organisms such as diatoms and fine filaments of cells that form blanket weed in ponds, to the great range of seaweeds that abound around the coast. The vast majority are never fossilised as complete organisms, but some, like diatoms, have tests of silica, and others secrete carbonate and are found as fossils. Many algae produce tough organic-walled cysts during the reproduction process, and these are recovered in palynological preparations.

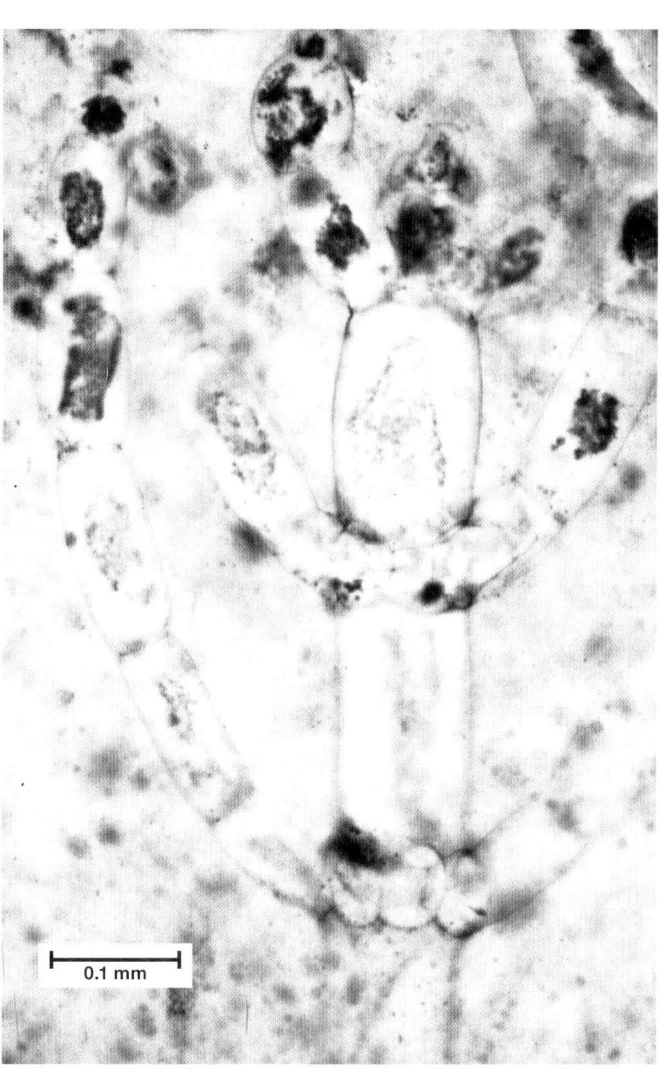

3 *Palaeonitella* preserved in Rhynie chert.

3

Palaeonitella / Alga / Devonian

Division	**Charophyta**
Class	**Charophyceae**
Gen et sp	***Palaeonitella cranii***
Locality	**Rhynie, Aberdeenshire**
Age	**Early Devonian**
Stratigraphy	**Rhynie Cherts Unit, Dryden Flags Fm., Lower Old Red Sandstone**

The charophyte algae are familiar inhabitants of ponds and streams at the present day and are known as stonewort. There are about 30 living British species, but *Palaeonitella* is the oldest fossil species in which the whole form of the alga is known. It consists of a stem made up of a string of single cells, from which whorls of branches emerge; the branches also subdivide as in the modern genus *Nitella*. Thus when the fossil was discovered it was given the name *Palaeonitella* ('ancient *Nitella*') by Kidston and Lang (1921). The specific name honours the Revd Cran of Skene, Aberdeenshire who discovered the specimens and gave them to Kidston and Lang to describe.

Palaeonitella was very delicate, and it is fortunate that it was rapidly silicified in the waters of the Early Devonian hot-springs of Rhynie (*see* Plants section). Apart from knowledge on the general form of the plant, the male and female reproductive organs were described recently by Kelman *et al.* (2004), making this the oldest charophyte of which the full life cycle is preserved. Normally only the calcified female reproductive organs (gyrogonites) of this group of algae are preserved in the geological record.

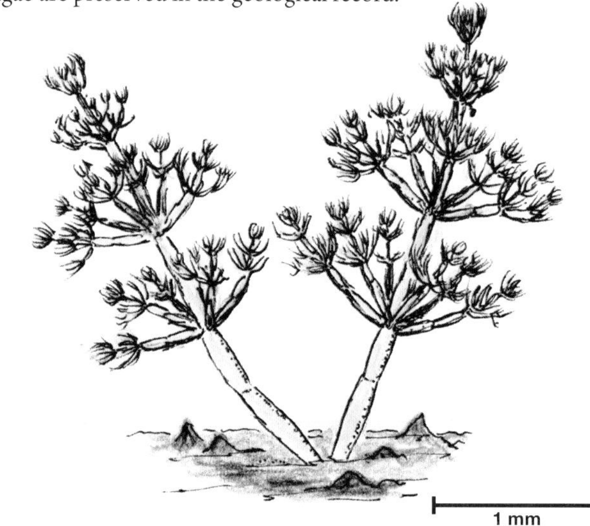

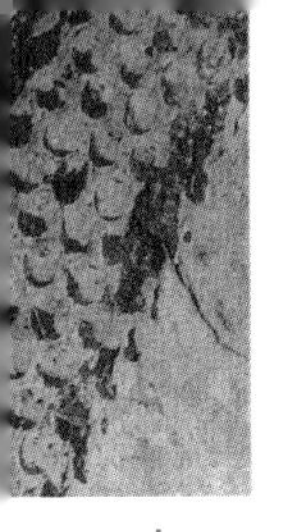

PLANTS

There are several well-known fossil floras found in Scotland, throwing light on the development of the land vegetation through geological time, and on the changing climate that Scotland has experienced.

The flora of the Early Devonian Rhynie and Windyfield cherts in Aberdeenshire is of world importance, recording the plants of a time when only small primitive forms had colonised wetter areas of the land. All the Rhynie plants were less than 50cm high. They are wonderfully preserved in 3D within chert (a silica rock similar to flint). The plants grew near a fossil hot spring that deposited silica, and mineralised or petrified the plants where they grew. The plants have to be studied in microscope sections, but superb cellular detail allows interpretations of plant structure not possible from crushed impressions. They were preserved before they were crushed and buried; some plants are even preserved with stems in growth position. The history of research into the Rhynie chert has been documented by Trewin (2004), and Fayers and Trewin (2004) provide an account of the biota and palaeoenvironment of hot spring pools and streams.

During the Carboniferous Scotland lay close to the Equator; tropical forests covered vast areas, and vegetation had now invaded higher areas of land. The rotted remains of the plants produced the coal that fuelled the industrial revolution in Scotland and is still exploited in opencast mines, although deep mines have all closed. The flora of the forest, with giant horsetails (*Calamites*) and trees such as *Lepidodendron,* has been extensively studied, revealing flora and conditions very different from modern tropical forests. Cleal and Thomas (1994) have provided an excellent guide to the plants of the Coal Measures and the conditions in which they lived.

Jurassic floras are represented by fossils found in shales in the Brora–Helmsdale region of Sutherland. The plants were deposited in the sea, but close to the Scottish landmass. Plants were transported from the land by river floods, and in the sea they became waterlogged and sank to the bottom to be preserved along with the marine fauna that lived in the nearshore area. Marie Stopes (1907) published a paper on the fossil plants from Brora, and her work and interaction with others in the Brora area is well documented by Falcon-Lang (2008). Stopes produced several palaeobotanical papers and also a textbook of palaeobotany (Stopes 1910), but she is better known for her role in the emancipication of women.

In Early Tertiary times there was extensive volcanic activity in western Scotland, with major volcanoes pouring out lava and ash from centres in Skye, Rum, Mull and Ardnamurchan. Plants that colonised the area are found preserved in sandy deposits between the lava flows, particularly on Skye and Mull (Boulter and Kvaček 1989). The plants reveal that Scotland enjoyed a warm climate with a variety of broad-leaved trees.

A few examples from the Scottish floras are illustrated below.

4

Parka / Plant / Devonian

Division	**May be allied to Mosses – non-vascular plant**
Gen et sp	***Parka decipiens***
Locality	**Aberlemno, near Forfar, Angus**
Age	**Early Devonian**
Stratigraphy	**Arbuthnott Group, Lower Old Red Sandstone**

Parka was a strange early land plant that seems to have grown as a thin sheet over the surface of sediment. It is classed as a non-vascular plant and does not fit easily in any modern classification. It has some features of mosses and algae. The upper surface bore rounded spore capsules that give it a spotted appearance. A likely habitat is the bare sediment surfaces left following river floods. It is very common in Lower Old Red Sandstone sedimentary rocks deposited by rivers and floods in the Forfar area, and occurs in both sandstones and shales. The organic material is generally black, with carbon preserved when it is in grey shales, but it is frequently oxidised and red when seen in sandstones.

This fossil has an interesting history. It was first mentioned by Fleming (1830), who compared *Parka* specimens to flattened raspberries, but concluded that they were part of the reproductive parts of a type of extinct rush. Hugh Miller described *Parka* from Carmyllie in *The Old Red Sandstone* (2nd edn 1842) suggesting dried up frogs' eggs, and local quarrymen also called them puddock (frog) spawn. This interpretation was supported by Mantell (1852).

Lyell (1841), in the second edition of *Elements of Geology*, considered that *Parka* might be gastropod eggs, and later they were regarded as eggs of the eurypterid *Pterygotus* by Huxley and Salter (1859), but the original idea probably came from James Powrie. It was suggested that a single layer of eggs was held in a connecting membrane. This theory was also followed by Murchison in 'Siluria' (4th edn 1867), and then adopted by Lyell. An interesting specimen of *Parka* from the Kinnaird Collection in Dundee Museum still retains a label stating 'ovisacs of *Pterygotus anglicus*'.

However, Hugh Miller had adopted a plant origin for *Parka* by 1855 (Miller 1857), but it was Dawson and Penhallow (1891) who clearly demonstrated that the tissue was that of a plant, and Don and Hickling (1917) confirmed this opinion when they extracted and figured plant spores from *Parka*. The history of study of *Parka* has been documented by Niklas (1976) and Hemsley (1990).

Thus *Parka* had the attention of many of the great men of nineteenth century geology, but eventually yielded its secrets to those who took time to examine the specimens in detail and discover facts. However, it still defies classification. There can be few fossils that have attracted the attention of such a string of famous geological names – even more remarkable for such a lowly plant.

4 *Parka*, in shale from the Lower Old Red Sandstone.

5

Rhynia / Plant / Devonian

Division	**Rhyniophyta**
Class	**Rhyniopsida**
Gen et sp	***Rhynia gwynne-vaughanii***
Locality	**Rhynie, Aberdeenshire**
Age	**Early Devonian**
Stratigraphy	**Rhynie Cherts Unit, Dryden Flags Fm., Lower Old Red Sandstone**

The Rhynie chert was originally deposited as silica sinter around hot springs similar to those in Yellowstone National Park, USA. Plants that grew on the land surface around the hot springs were preserved in silica when hot spring waters invaded areas with plant growth. The result was the preservation of plants and arthropods in superb detail, allowing a glimpse of an Early Devonian ecosystem (Trewin 1994).

The spore-bearing plants *Rhynia* and *Asteroxylon* are featured here and are two of the seven spore-bearing plants that have been described from the chert. The illustrated slab of chert with *Rhynia* also includes a few *Agalophyton* axes and a cross-section of a sporangium. Details of the plants have been reviewed by Edwards (2004). The alga *Palaeonitella,* and arthropod *Palaeocharinus* are also included in this book, and the general biota is illustrated on the Web (Trewin *et al.*, 2001).

In the best material *Rhynia* is preserved in three dimensions with excellent cellular preservation. The reconstruction of *Rhynia,* and other plants, is based on microscope work using thin sections of the chert. *Rhynia* was a very simple primitive plant having naked stems up to 20cm high arising from creeping rhizomes. Tufts of rhizoids on the rhizomes aided the takeup of water and helped fix the plant on the sandy land surface. The plant reproduced from spores that were held in a sporangium at the tip of an upright stem. The fact that stomata covered the axes shows that this was a land plant. *Rhynia* was found by William Mackie, and first described by Kidston and Lang (1917). The specific name celebrates the botanist Gwynne-Vaughan who was to have collaborated with Kidston on describing the Rhynie flora, but died early in the project. Kidston and Lang described two species of *Rhynia,* but their *Rhynia major* has been redescribed and named *Aglaophyton* by Edwards (1986). *Rhynia* is probably the most common plant in the chert, and is frequently found on its own, implying that it was an initial coloniser of sandy surfaces.

5a Polished chert with *Rhynia* axes in growth position. Larger axes at left and elliptical sporangium are *Aglaophyton*. Fine horizontal banding in pale chert is bacterial mats.

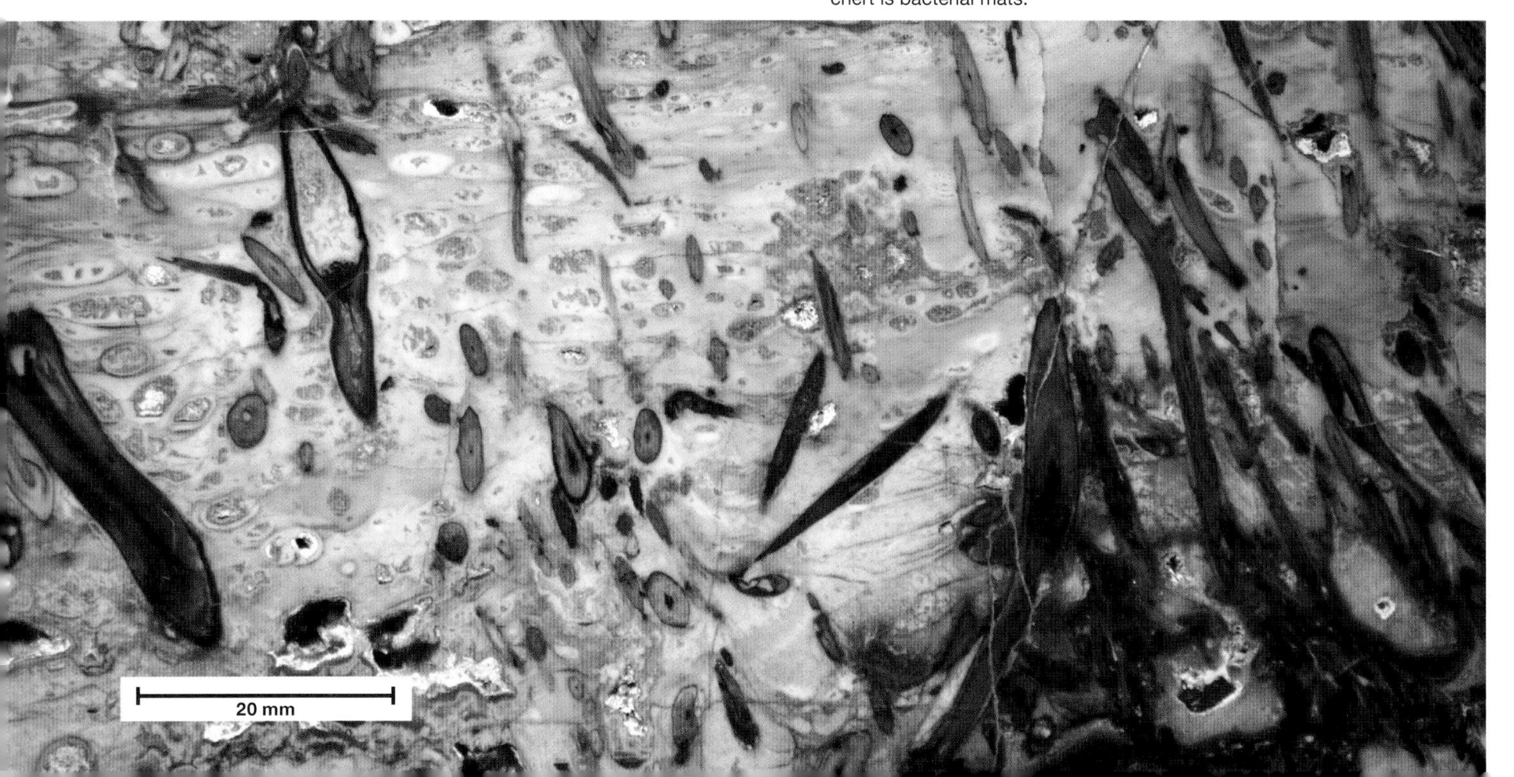

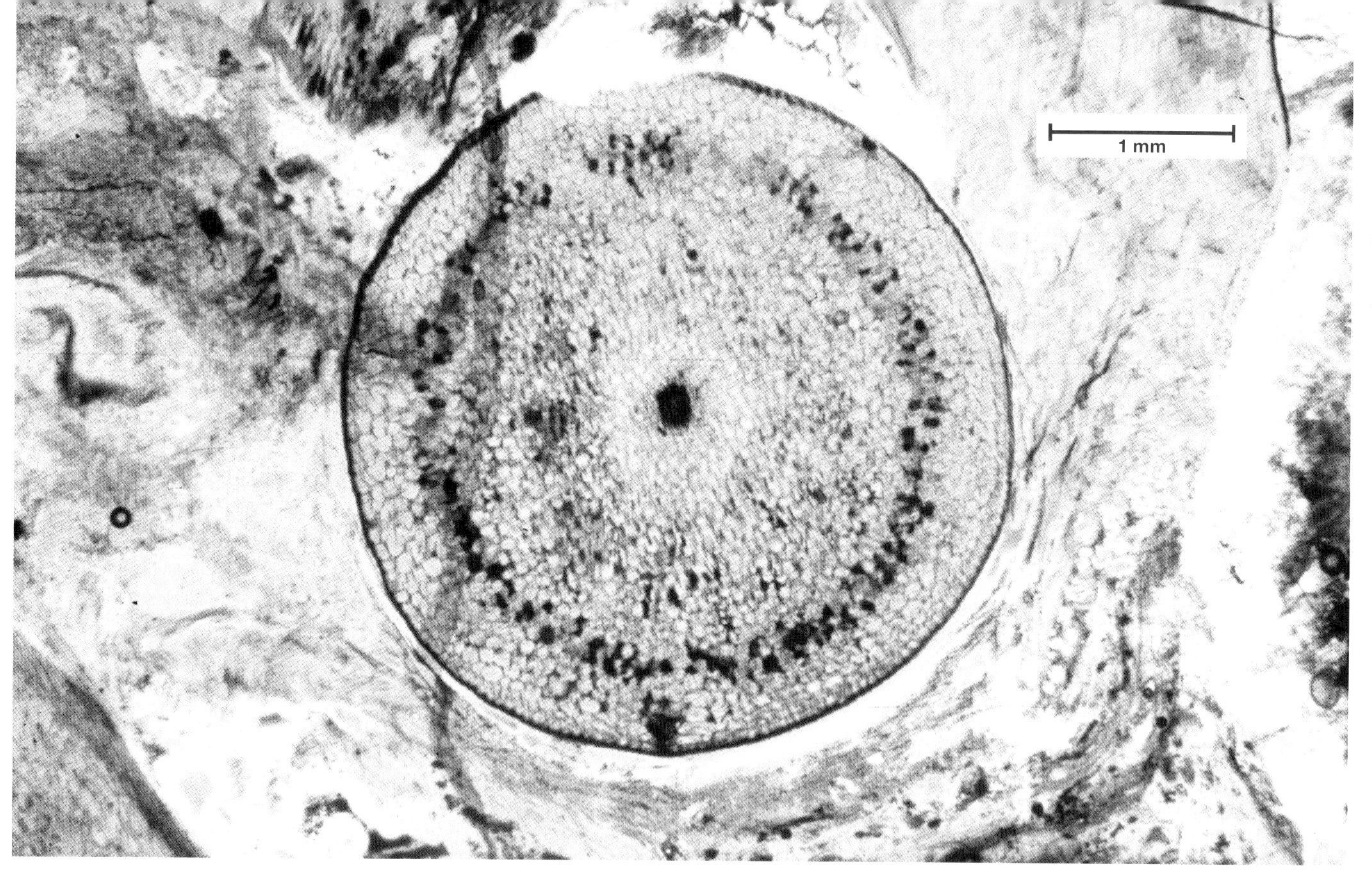

5b *Rhynia* axis in cross-section.

5c Model of *Rhynia*.

6

Asteroxylon / Plant / Devonian

Division	Lycophyta
Class	Lycopsida
Gen et sp	*Asteroxylon mackei*
Locality	Rhynie, Aberdeenshire
Age	Early Devonian
Stratigraphy	Rhynie Cherts Unit, Dryden Flags Fm., Lower Old Red Sandstone

Asteroxylon was a larger and more complex plant than *Rhynia.* The upright stems were clothed in small fleshy bracts (not true leaves), and the sporangia were kidney-shaped and grouped near the tip of the stem. This was the largest plant at Rhynie,

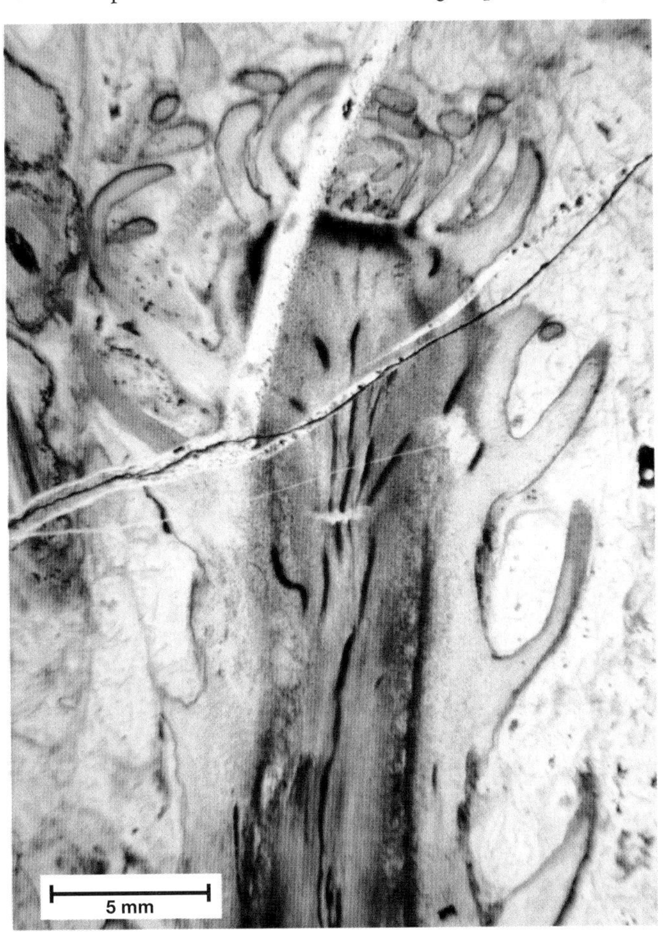

6a *Asteroxylon*, longitudinal section of axis tip.

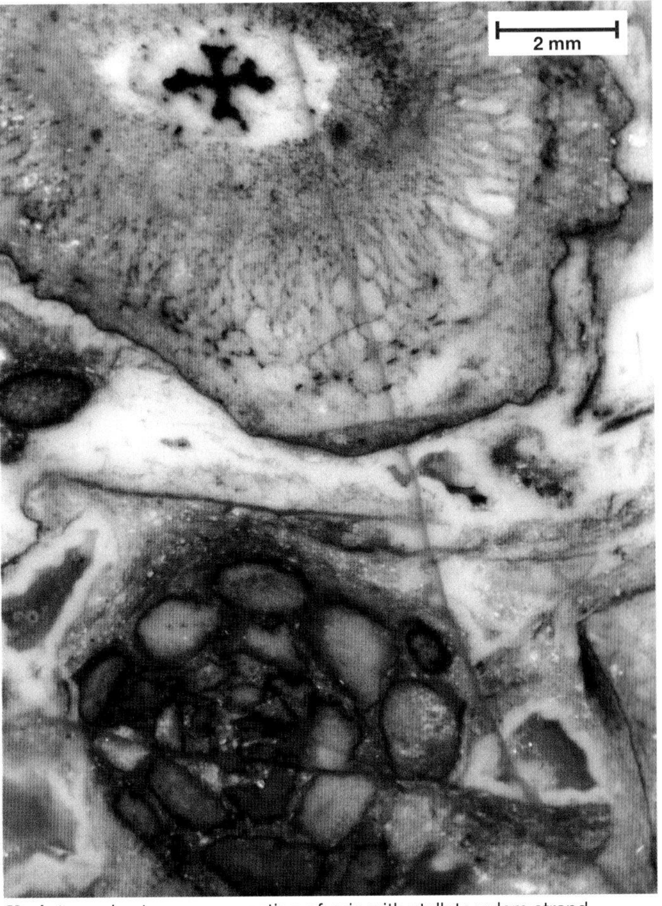

6b *Asteroxylon* transverse section of axis with stellate xylem strand (top), and transverse section of axis tip (below).

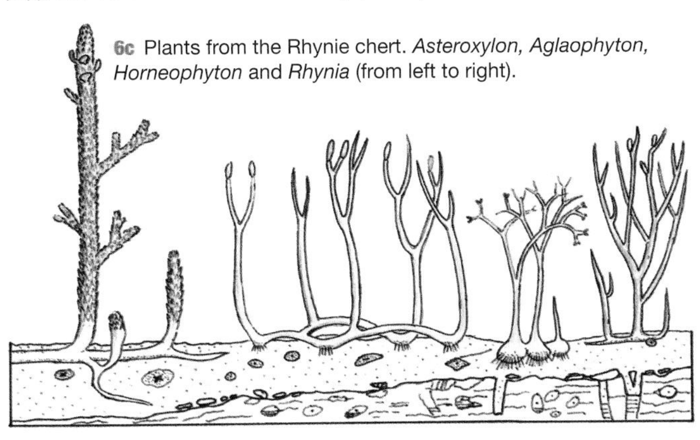

6c Plants from the Rhynie chert. *Asteroxylon, Aglaophyton, Horneophyton* and *Rhynia* (from left to right).

but still only 50 cm tall. A complex lobed xylem strand within the stems gave the added strength required to support the tall stems. A strong, branching rhizome system penetrated the plant litter and sandy soil in which the plant grew, and gave added support. *Asteroxylon* is normally found associated with several of the other Rhynie chert plants, and seems to have been part of a more mature plant community, rather than an early coloniser of bare surfaces. There is a strong similarity between *Asteroxylon* and the modern lycopsid *Hupertzia selago*, which can be found commonly on Scottish mountains.

7

Stigmaria / Plant / Carboniferous

Division	**Lycophyta**
Class	**Lycopsida**
Gen et sp	**Stigmaria**
Locality	**Victoria Park, Glasgow**
Age	**Carboniferous**
Stratigraphy	**Limestone Coal Formation**

Stigmaria is a general name given to root casts of lycopod trees, or giant club-mosses of the Carboniferous. In Victoria Park, Glasgow, eleven tree stumps and branching roots cast in sandstone have been preserved as the 'Fossil Grove'. The stumps are preserved as they grew in the Carboniferous forest, and give an excellent impression of the spacing of the trees, estimated at 4500 per square kilometre. The tree stumps were discovered in 1887 and incorporated into the park. This is one of the earliest examples of geological conservation. More usually tree stumps are not preserved, and only sand-filled casts of the roots remain as fossils.

7 Sandstone casts of *Lepidodendron* tree stumps with attached stigmaria (root casts).

8

Lepidodendron / Plant / Carboniferous

Division	**Lycophyta**
Class	**Lycopsida**
Gen et sp	***Lepidodendron veltheimianum***
Locality	**Burdiehouse, Midlothian**
Age	**Carboniferous**
Stratigraphy	**Calciferous Sandstone Series**

The lycopod *Lepidodendron* grew as tall trees in the Carboniferous forests. The distinctive bark pattern of diamond-shaped leaf scars is the part of the tree most often found as a fossil. The mature tree had a tall, straight trunk capped by much-divided branches, the whole effect being similar to that of a palm. Rarely the tips of the branches may be found with slender leaves still in place. This tree was a major component of the wet swampy forests of the late Carboniferous where waterlogged vegetation accumulated as peat, eventually to be transformed to coal with deep burial. Numerous specific names have been applied to *Lepidodendron*; the name given above is merely repeated from the label on the specimen; this is not a guarantee that an expert might consider the name to be correct today!

The example illustrated is from the Hugh Miller Collection, in the National Museums, Scotland collection, and illustrates the fact that Hugh Miller continued to add to his collection following his move from his home town of Cromarty to Edinburgh, where he was editor of *The Witness* newspaper.

8a Small branch of *Lepidodendron.*

8b Reconstruction of *Lepidodendron* tree, with detail of stigmaria.

9

Neuropteris / Plant / Carboniferous

Division	**Gymnospermophyta**
Class	**Cycadopsida**
Gen et sp	***Neuropteris loshii***
Locality	**Seafield Colliery, Fife**
Age	**Carboniferous, Westphalian B**
Stratigraphy	**Coal Measures Group, shales above Coxtool coal**

Neuropteris is the name given to distinctive lobed leaves that are frequently abundant in the mudstones of the Carboniferous coal swamps. The leaves were borne on branched fronds about a metre long that formed the crown of a tree about 5 m high, though some related forms were more than twice that size. These primitive 'seed ferns' produced seeds that were naked, rather than being enclosed in an ovary, as in later angiosperms. They grew along the banks of rivers; thus leaves and fronds that fell into the water had a good chance of burial in mud and preservation as fossils.

9 *Neuropteris* leaf.

10

Calamites / Plant / Carboniferous

Division	Sphenophyta
Class	Equisetopsida
Gen et sp	*Calamites cistii*
Locality	Seafield Colliery, Fife
Age	Carboniferous, Westphalian B
Stratigraphy	Coal Measures Group, shales above Coxtool coal

Calamites looked like a giant version of the modern *Equisetum* (mare's-tails) in having a ribbed stem with regularly spaced whorls of narrow leaves or branches. However, whilst modern *Equisetum* are seldom over a metre high and die back in winter, *Calamites* continued growing from season to season, producing tree-sized plants up to 20 m high. They grew in forests with taller trees such as *Lepidodendron*, and their fossils are generally sand-filled casts of hollow stems, or flattened stems, as seen in the illustration.

10a *Calamites* stems.

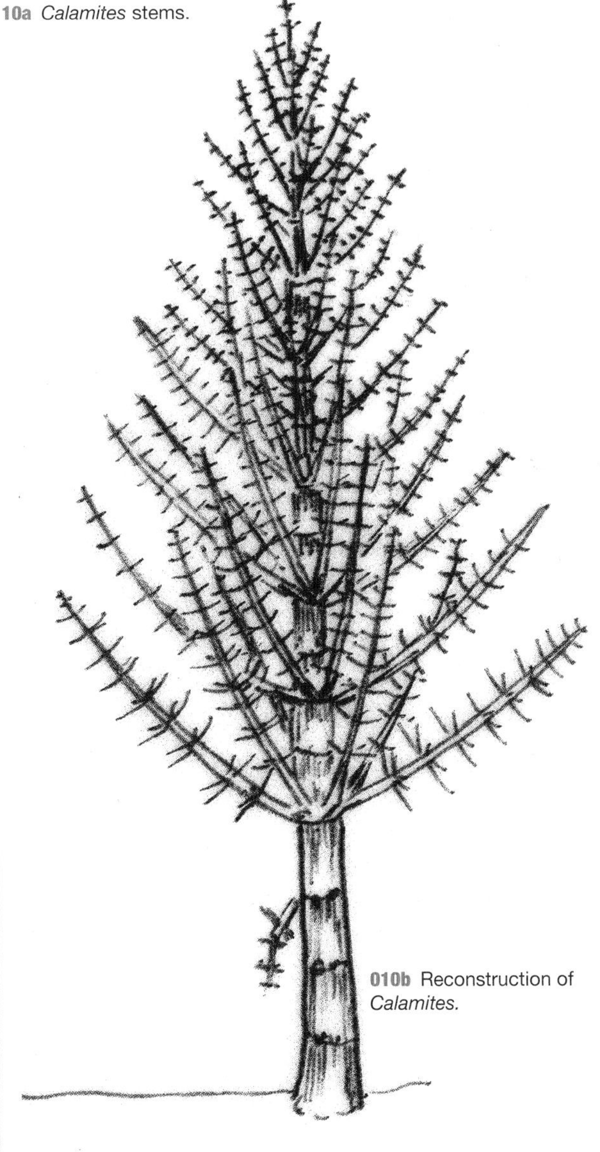

010b Reconstruction of *Calamites*.

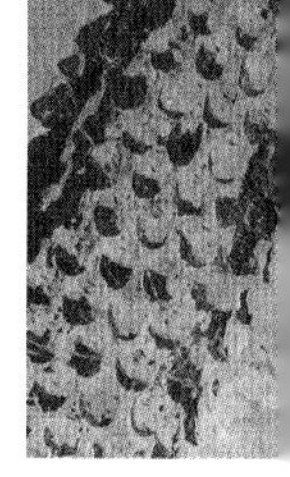

11

Mariopteris / Plant / Carboniferous

Division	**Pteridophyta**
Class	**Lagenostomopsida**
Gen et sp	***Mariopteris latifolia***
Locality	**Musselburgh**
Age	**Late Carboniferous**
Stratigraphy	**Coal Measures**

Mariopteris was an early seed plant, and it probably grew as a liana, climbing up other trees to rapidly reach light in the canopy of the Carboniferous forests. It is difficult to identify these and many other Carboniferous plants because only small fragments are generally available, and plants can show great variation in leaf or frond shape from one part of a tree to another. Carboniferous plants were frequently named on minor morphological differences, and thus the names cannot be considered to be botanical genera, but only 'form genera', several of which might occur on the same plant.

This specimen is from the Hugh Miller Collection held in National Museums, Scotland in Edinburgh.

11 *Mariopteris* frond.

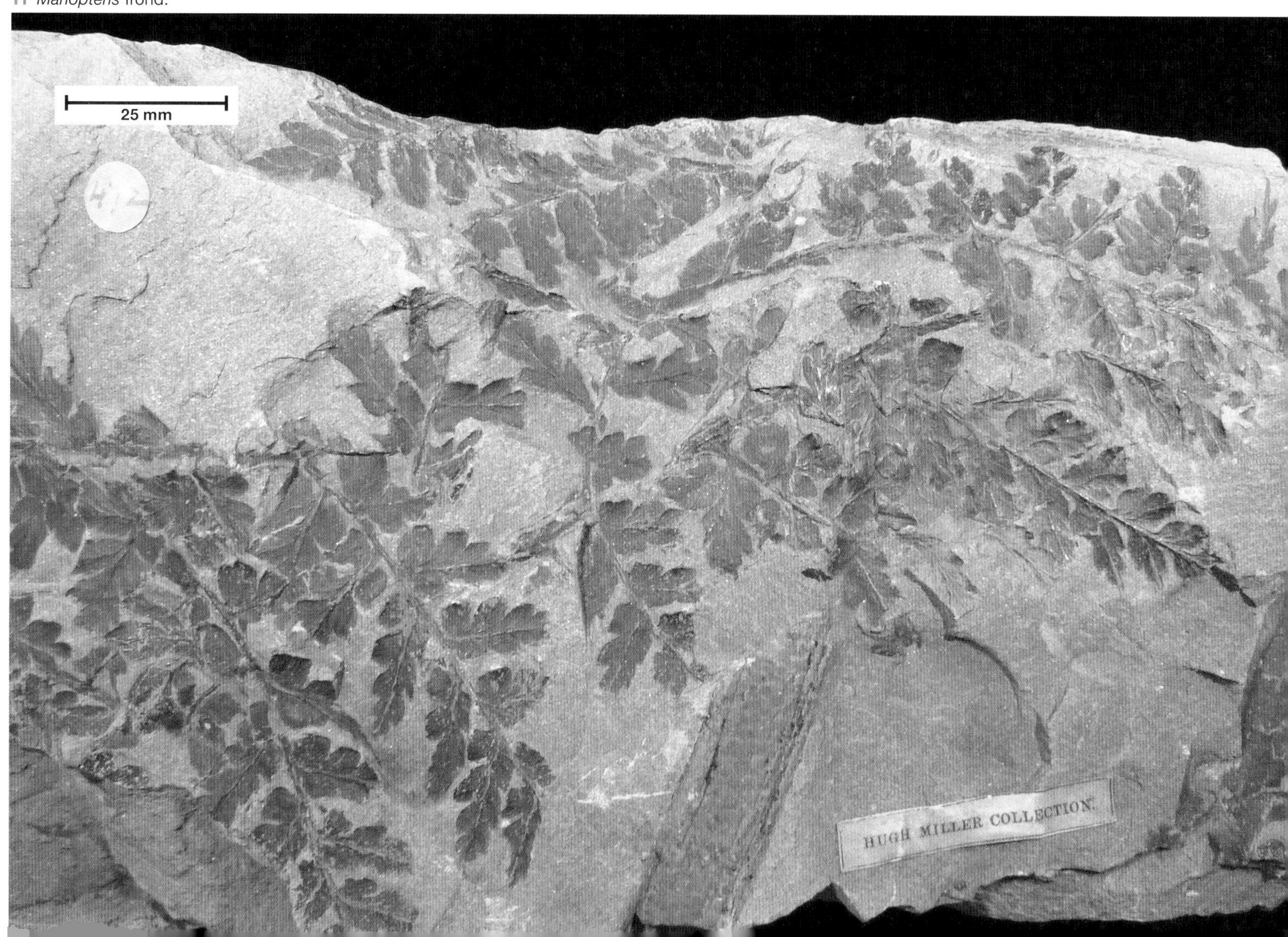

12

Ptilophyllum / Plant / Jurassic

Division	**Gymnosoophyta**
Class	**Cycadopsida**
Gen et sp	***Ptilophyllum* cf. *pectinoides***
Locality	**Lothbeg Point, near Brora, Sutherland**
Age	**Late Jurassic, Kimmeridgian**
Stratigraphy	**Helmsdale Boulder Beds**

This is part of a frond of *Ptilophyllum* found in a loose block of laminated limestone on the shore at Lothbeg Point. The rock was eroded by the sea from shales within the Helmsdale Boulder Beds and cast up on the beach. *Ptilophyllum* was a bennettitalian, rather similar to a cycad in general form, and is part of a varied flora of over 30 species of plants that have been found at the locality. Pteridosperms, ferns, cycads and conifers are well represented and reflect plant life on a swampy coastal delta bordering the Scottish Jurassic land area. The plant remains were washed into the sea during floods, became waterlogged, and sank. They are preserved in marine shales along with ammonites and belemnites. The fossil plants of this locality have been described by Van der Burgh and Van Conijnenberg-Van Cittert (1984).

012 *Ptillophyllum* frond.

13

Zamia / Plant / Jurassic

Division	**Gymnosoophyta**
Class	**Gnetopsida**
Order	**Bennettitales**
Gen et sp	***Zamia* sp.**
Locality	**Helmsdale, Sutherland**
Age	**Late Jurassic, Kimmeridgian**
Stratigraphy	**Helmsdale Boulder Beds**

This frond of *Zamia* is from the Hugh Miller Collection in the Royal Museums, Scotland. It was illustrated by Hugh Miller in *The Testimony of the Rocks* (1857, Fig. 137, p. 479). It is not always easy to relate engraved figures in old publications to actual specimens, since the author first did a drawing, and this had to be copied by the engraver. Authors were prone to exaggerate features they wanted to stress, and omit features that they considered unimportant. Hugh Miller's illustration is reproduced here with a photo of the specimen. *Zamia* is thought to belong to the bennettitalians, which were similar to cycads.

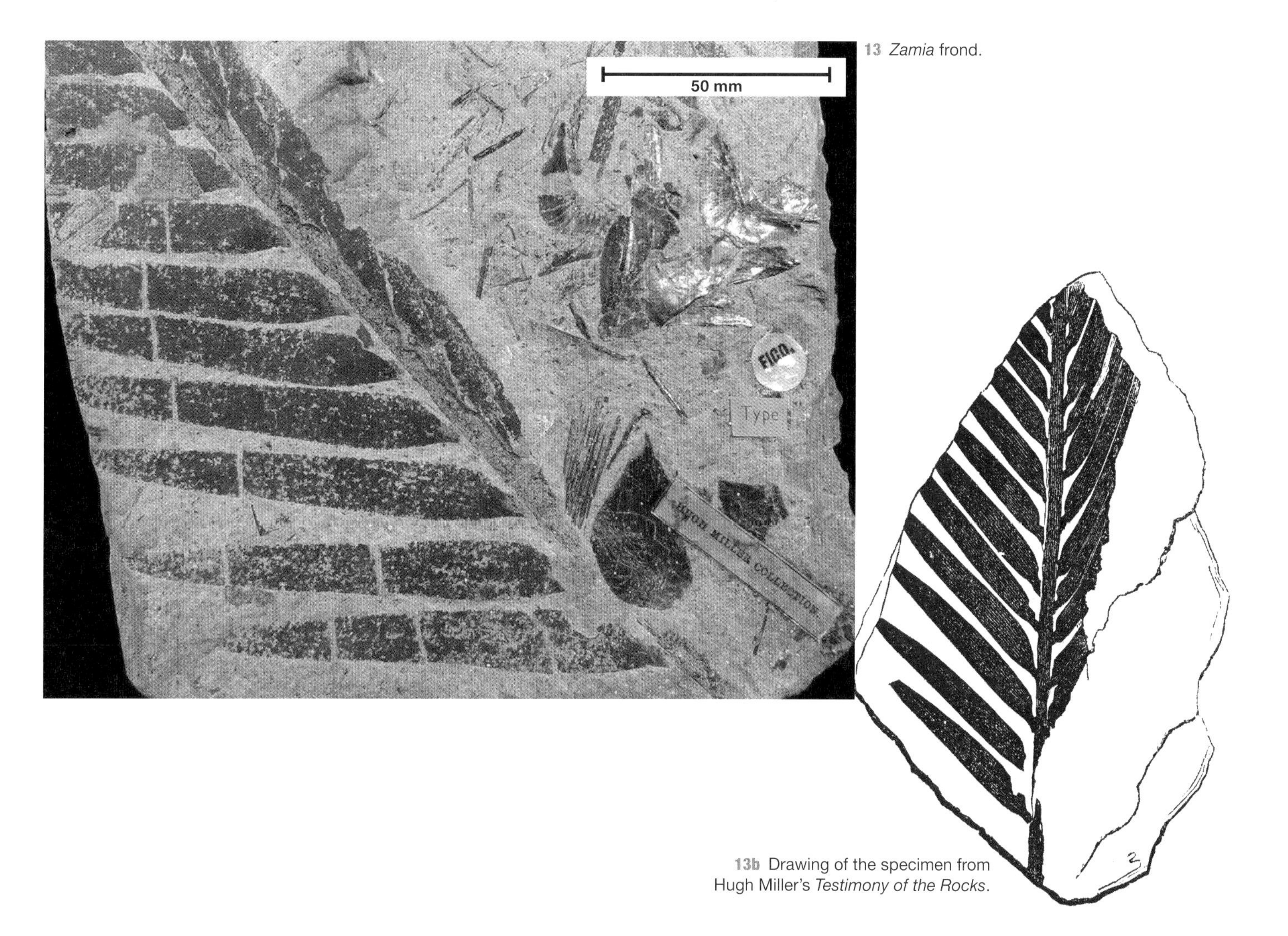

13 *Zamia* frond.

13b Drawing of the specimen from Hugh Miller's *Testimony of the Rocks*.

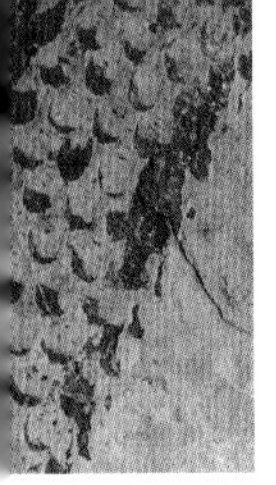

14

Macculloch's Tree / Plant / Palaeocene

Division	**Gymnospermophyta**
Class	**Pinopsida**
Gen et sp	***Taxodioxylon***
Locality	**Ardmeanach, Isle of Mull**
Age	**Palaeocene**
Stratigraphy	**Staffa Lava Formation**

Macculloch's tree is named after John Macculloch, who described and figured this famous fossil in his *Description of the Western Isles of Scotland* (1819). Macculloch was a pioneer of Scottish Geology and produced the first detailed geological map of the whole of Scotland in 1836. Unfortunately he did not live to see the map published as he died, aged 61, following a carriage accident whilst on his honeymoon in Cornwall in 1835.

The tree is preserved in a vertical position within a lava flow displaying excellent columnar jointing, which fans out around the tree. Much of the 'tree' is a hollow cast that filled with volcanic

14a Macculloch's tree, encased in lava.

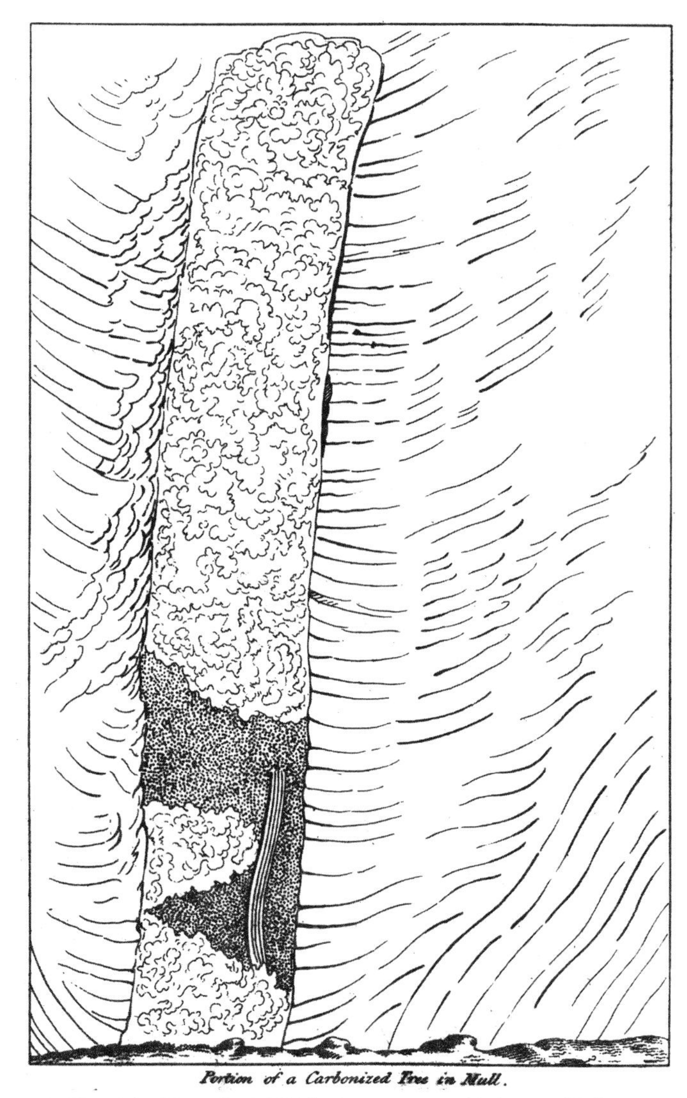

14b Sketch by Macculloch (1819) of tree preserved in lava on Mull.

debris after the tree was burnt and decayed, but some woody tissue remained at the base of the tree. This was a large coniferous tree some five feet in diameter, and estimated by Macculloch to be about 50 feet high. Wood of *Taxodioxylon* closely resembles *Sequoia*, thus one can envisage a forest of redwoods growing on Mull before the fatal eruption. The lava flowed rapidly around the tree, burying much of the trunk. The exposed top would have been burnt before the upper part of the flow solidified. This was an extensive lava flow that can be correlated with the flow forming the columns of Fingal's Cave on Staffa. The tree is in a remote locality, and it is a long and rough walk to the locality.

There are other similar trees preserved in lava on Mull, notably the Carsaig Tree illustrated by Brian Bell in *The Geology of Scotland* (2002, 4th edn, Plate 26). It is clear that the lava flow destroyed a large area of mature forest; it seems to have flowed very rapidly into hollows where the lava became ponded, so building rapidly to a considerable depth and encasing the larger trees.

15

Platanites / Plant / Palaeocene

Division	**Angiospermae**
Class	**Magnoliopsida**
Gen et sp	***Platanites hebridicus***
Locality	**Ardtun, Isle of Mull**
Age	**Palaeocene**
Stratigraphy	**Ardtun Leaf Beds, Staffa Lava Formation**

The Ardtun Leaf Beds (Boulter and Kvaček 1989) provide an important glimpse into the plants that lived in western Scotland at the time of major volcanicity at the Palaeocene/Eocene boundary about 55Ma ago. *Platanites* (the plane tree) is chosen as a representative of the broad-leaved deciduous trees that flourished alongside conifers in the warm climate. Other plants recognised include ferns, cycads, conifers such as *Metasequoia*, and *Ginkgo*. Mixed woodlands flourished in soils derived from the weathering of basalt lavas. Scotland was probably covered in rich woodland at this time, but only the fossils at Ardtun, and some localities on Skye, remain to provide the fossil evidence.

14c Mucculloch's tree, detail of bark.

15 *Platanites* leaf from Ardtun, Mull.

SPONGES

Sponges are simple animals and consist of only a few types of cells. The cells deposit, and are supported by, spicules made of carbonate, silica or spongin. The basic plan of a sponge comprises an outer wall with small openings; beneath the surface these openings connect to larger chambers, and hence to a central cavity. Water is drawn in through the surface of the sponge by cells in the chambers that each have a whip-like flagellum. The cells work in unison to create a current of water using the flagella. The sponge extracts food particles from the water, which is then expelled from the central cavity. Thus a current of water continuously passes through the sponge. The individual cells of sponges closely resemble protozoa (single-celled organisms) but they can organise themselves and work in unison. Sponges take on many shapes, and are generally fixed to the seabed, cemented to rocks or rooted in sediment. There is no nervous system or gut, and the cells are not organised into organs as in a metazoan. Whilst there exists a wide variety of sponges, they have never evolved into metazoans.

Sponges range from the Cambrian to the present day, and at some periods were so abundant that sponge spicules can form a sedimentary rock known as a spiculite.

16

Ventriculites / Sponge / Cretaceous

Phylum	**Porifera**
Class	**Hexactinellida**
Gen et sp	***Ventriculites***
Locality	**Moreseat, near Peterhead, Aberdeenshire**
Age	**Cretaceous (but found in a pebble from Pliocene gravel)**
Stratigraphy	**Buchan Ridge Gravel Formation, (derived from Cretaceous chalk)**

Ventriculites is a common vase-shaped sponge of the Cretaceous, and was abundant in the seas where chalk was deposited. This specimen is preserved within a nodule of flint that was derived from the chalk. On breaking the flint the sponge has neatly broken out, and reveals some of the structure. The tiny surface pores are visible on the impression left in the flint, and larger circular canals are seen on the conical sponge. The central cavity of the sponge is filled with flint.

This sponge had spicules made of opaline silica, and it is thought that this type of sponge spicule provided the silica that was deposited in the chalk as flint. The amorphous opaline sponge spicules dissolved in pore water within the chalk, and the silica was redeposited in a more stable crystalline form as flint nodules. In this case the flint formed around a buried sponge.

16a Conical *Ventriculites* sponge and cast in flint.

The locality of Moreseat is unusual because flint-rich gravel occurs on moorland hilltops, but there is no chalk in the area to supply the flints. There is also some Lower Cretaceous sandstone in the same area that may have been deposited near a Cretaceous shoreline. William Ferguson (1850) recorded chalk flints and 'Greensand' fossils from Moreseat, and there is considerable literature on these flint gravels of the Buchan Ridge, summarised by Hall (1993). There are various opinions on the deposition of the flint gravels with rivers, glaciation and marine gravels all proposed, as discussed by Bridgland *et al.* (1997). It seems probable that Upper Cretaceous chalk with flints once covered this corner of NE Scotland, and that the soft chalk weathered away to leave the resistant flints, which were reworked into gravels. These flint gravels are thought to be Pliocene or earlier in age, and are overlain by till from the Pleistocene ice ages. At the Den of Boddam there is evidence from radiocarbon dates that the flints were quarried in the late Neolithic for tool manufacture between roughly 3200 and 2200 BC (data in Bridgland *et al.* 1997).

16b Detail of outer surface of *Ventriculites*.

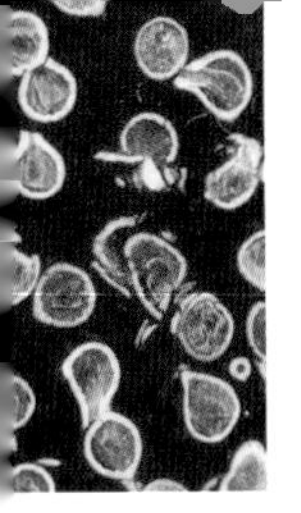

CORALS

Corals are generally associated with reefs, lagoons and tropical conditions. The Great Barrier Reef of Australia is a spectacular modern example. The corals in modern tropical reefs require clean, clear, warm, shallow water to survive. Sunlight is essential to nourish micro-organisms that live within the coral tissue, and without which the coral cannot survive (a symbiotic relationship). There are also deep-water coral reefs, some in the North Atlantic on the edge of the continental shelf. These reefs have very low diversity of species, and the corals can survive in the cold and dark conditions. The deep-water corals do not depend on a symbiotic relationship with photosynthetic micro-organisms.

Modern corals belong to the Order Scleractinia, and these are the main reef-building corals of the Mesozoic to the present day. In the Palaeozoic the Orders Rugosa and Tabulata provided the reef corals, and they built reefs that differed from modern reefs, generally forming as flat or domed structures that did not build up into the surf zone. They are also associated with more muddy sediments, indicating that they lived in lower energy conditions.

Common fossil examples of the three major groups of corals are illustrated, with brief details of their characteristic features. Also included is *Conularia*, a member of an extinct group probably rlated to sea anemones or jellyfish.

17a Colony of *Syringopora*.

17

Syringopora / Coral / Carboniferous

Phylum	Cnidaria
Class	Anthozoa
Order	Tabulata
Gen et sp	*Syringopora*
Localities	New Cumnock, Ayrshire (17a,b); Arbigland, Kirkudbrightshire (17c)
Age	Early Carboniferous
Stratigraphy	Lower Limestone Group

Syringopora is chosen as a common example of a tabulate coral; this is an extinct group, but was important in reef building in the Palaeozoic. They are distinguished on the basis of internal structures, having tabulae, and lacking the septa characterising the rugose corals of the time. The tabulae are horizontal or domed calcite plates, within the coral calyx. Tabulate corals were colonial, with an individual animal at the end of each tube. The colony grew by the splitting of individual animals and the creation of more tubes. The colonies grew in a variety of shapes depending on sea-floor conditions, with bun-shaped colonies being common.

Tabulate and rugose corals lived together in warm shallow seas that periodically invaded the Midland Valley at times of high sea level in Carboniferous times. They occur in limestones with numerous brachiopods and crinoids.

17b Detail of *Syringopora* showing joins between tubes.

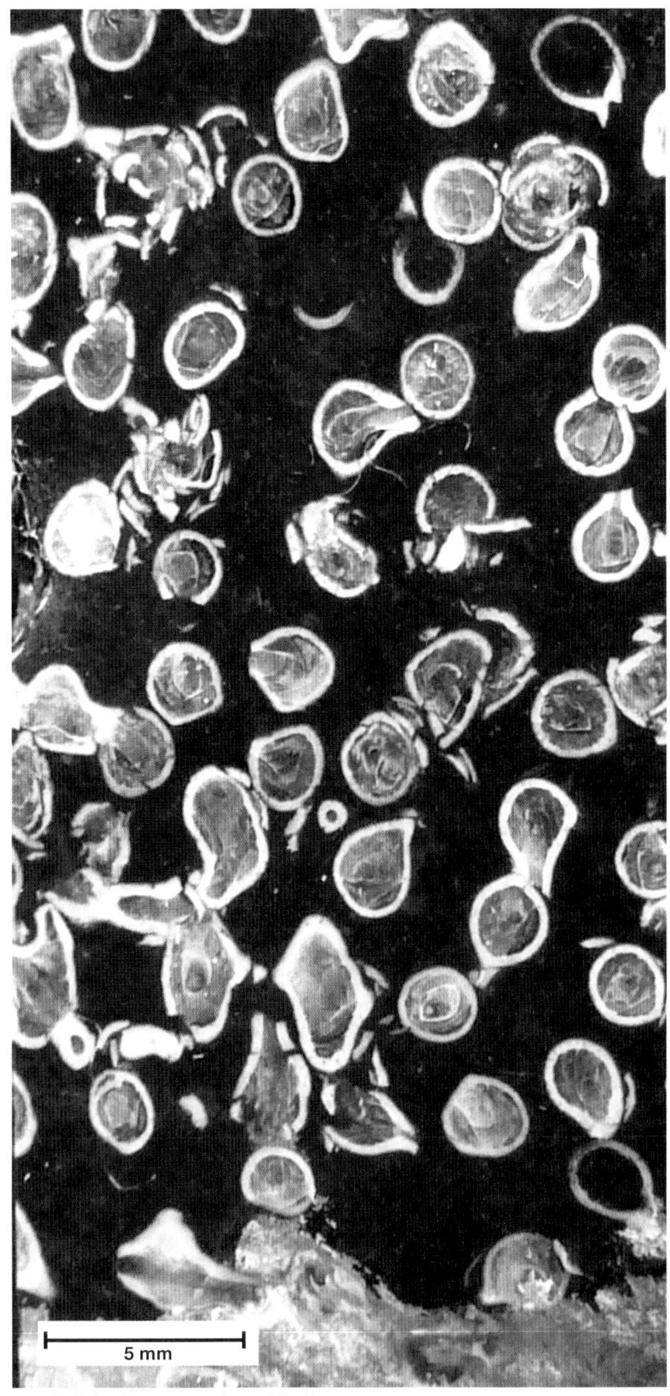

17c Polished cross-section of *Syringopora*.

18

Lithostrotion / Coral / Carboniferous

Phylum	**Cnidaria**
Class	**Anthozoa**
Order	**Rugosa**
Gen et sp	***Lithostrotion***
Locality	**Southerness, Dumfries**
Age	**Early Carboniferous**
Stratigraphy	**Lower Limestone Formation**

Lithostrotion is a common example of a colonial or compound rugose coral. The internal structure is more complex than within tabulate corals, with septa, dissepiments and tabulae forming a complex series of interlocking plates, giving strength to the colony and providing support to the individual coral animals. Different species are distinguished on the basis of the internal structure. Colonies are usually dome-shaped, and are found in limestones. The example illustrated shows the surface of part of a colony with the roughly hexagonal cups, each of which held a coral polyp. Sometimes they are found preserved in growth position in the rock, but they may also be overturned, giving evidence for ancient storm conditions breaking and transporting the colonies. At Southerness, Carboniferous limestones with corals were deposited in shallow seas close to the shore of a Carboniferous island.

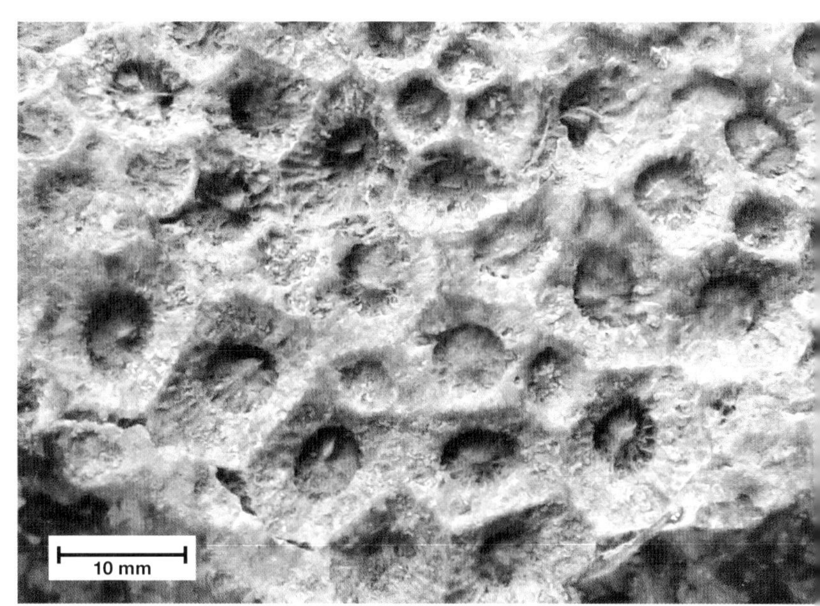

18 *Lithostrotion*, surface of colony.

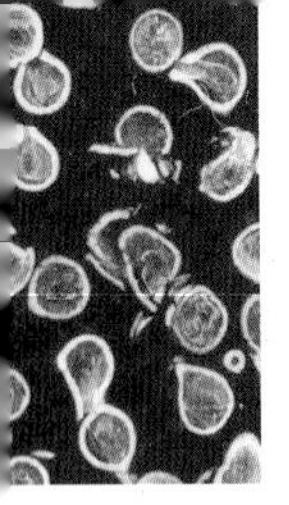

19

Koninckophyllum / Coral / Carboniferous

Phylum	**Cnidaria**
Class	**Anthozoa**
Order	**Rugosa**
Gen et sp	***Koninckophyllum***
Locality	**Barns Ness, East Lothian**
Age	**Early Carboniferous**
Stratigraphy	**Upper Longcraig Limestone, Lower Limestone Formation**

Koninkophyllum is a solitary rugose coral, meaning that a single coral animal built the irregular cup-like skeleton. It would have looked like a sea anemone sitting on an irregular cup. This is a common design for corals, and is found in modern reef corals, despite the fact that modern corals belong to a different class of the Cnidaria.

On the outside, the coral cup shows growth lines, and internally there is a complex arrangement of calcite plates. The details of the arrangement of these internal structures forms the basis for the classification of these fossils; hence they have to be studied in sections cut from the fossil. This is a case where it is not possible to be sure of an identification until the coral has been cut and the internal structure studied.

At Barns Ness one particular bed within the Upper Longcraig Limestone is crowded with these corals, which must have formed a carpet on the sea floor. The limestone was formed when sea level was high and shallow seas invaded a low-lying coast. When sea level fell, swamps and forest extended seawards, and coals were deposited. At one point in the Barns Ness section coal deposited as peat in a delta swamp can be seen resting directly on marine limestone with corals; clear evidence of changing relative sea level.

19 *Koninckophyllum*. Solitary corals in cut and polished limestone.

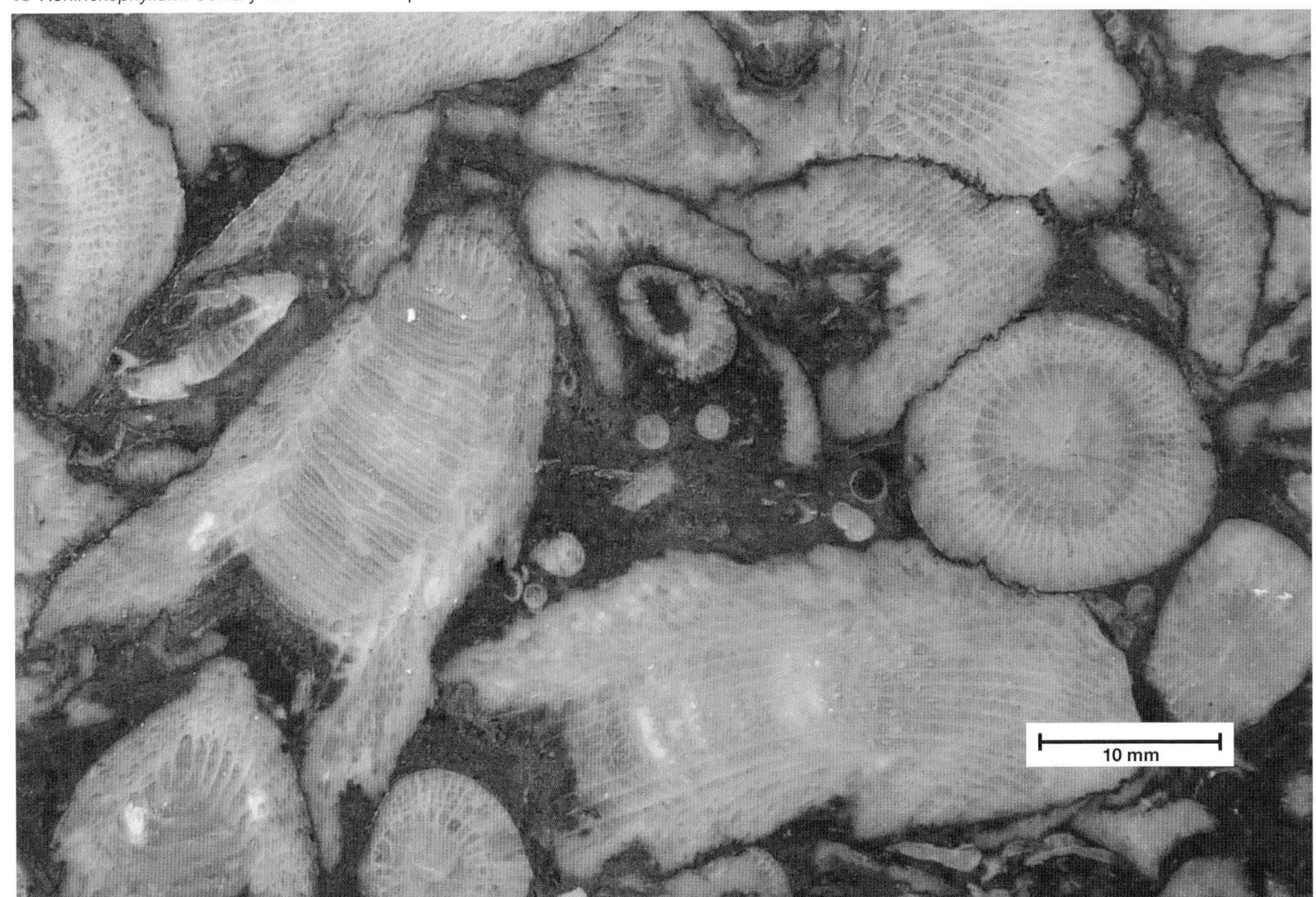

20

Isastraea / Coral / Jurassic

Phylum	**Cnidaria**
Class	**Anthozoa**
Order	**Scleractinia**
Gen et sp	***Isastraea oblonga***
Locality	**Helmsdale, Sutherland**
Age	**Late Jurassic, Kimmeridgian**
Stratigraphy	**Helmsdale Boulder Beds**

The coral *Isastraea* can be found on Helmsdale beach as specimens that range from pebbles to boulder-sized colonies. These Jurassic scleractinian coral colonies have been eroded from the Helmsdale Boulder Beds, which were deposited as rockfalls and debris flows from a submarine cliff that formed along the line of a major geological fault – the Helmsdale Fault. The general features of the Helmsdale Fault and the faunas of the Helmsdale Boulder Beds have been summarised by Hudson and Trewin (2002). The corals grew attached to rocks in shallow water, but were swept into deep water following earthquakes resulting from movements on the fault. Hence the coral colonies are now found in the Boulder Beds that form the rocks exposed on the beach. Along with the corals, bivalves, belemnites, ammonites and fossil wood are also present, together with rare bones of turtle, crocodile and *Plesiosaurus.* The field geology of the Helmsdale area has been described by MacDonald and Trewin (2009).

These colonial corals trapped food particles with their tentacles, and secreted carbonate to form the hard skeleton. Originally the colony would have been much lighter and porous, but all the spaces have been filled with calcite during the fossilisation process. Some colonies have borings that were made by bivalves and worms. In the Late Jurassic Scotland was closer to the Equator, at about the northern limit of coral growth, and enjoyed a warmer climate than at the present.

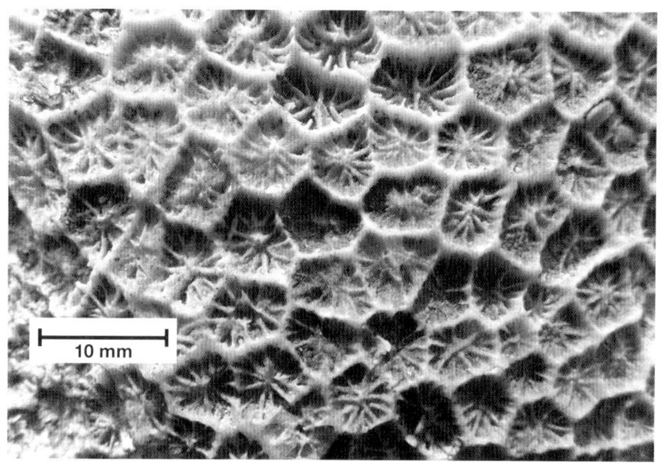

20b Detail of surface of *Isastraea* colony showing detail of individual corallites.

20a Large rounded colony of *Isastraea*.

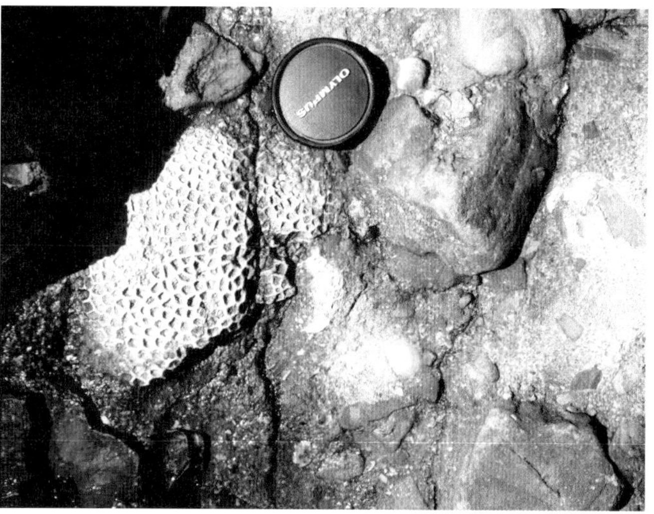

20c Coral colony in the Helmsdale Boulder Beds, Helmsdale.

21

Conularia /Jellyfish? / Carboniferous

Phylum	**Cnidaria**
Class	**Scyphozoa, Conulariida**
Gen et sp	***Conularia quadrisulcata***
Locality	**Walkmill Glen, Barrhead, Glasgow**
Age	**Carboniferous**
Stratigraphy	**Upper Limestone Formation**

Conularia is a fairly common fossil with a distinctive pyramidal shape. It is found in marine deposits in Scotland ranging from the Ordovician to Carboniferous, and has a worldwide range from late Cambrian to early Triassic time. The shell is chitinous and phosphatic and has a square cross-section, tapering to a point that appears to have been an attachment point, at least in early life. Each side is ornamented with chevron ridges meeting at a dividing line at the centre of each side. Some specimens show weak internal divisions, dividing the interior into four parts. A few specimens have four flaps at the end that could be used to close the shell.

A four-fold symmetry is seen in jellyfish, and it is thought that an animal similar to a jellyfish lived in the shell and extended tentacles to feed. A few exceptionally preserved specimens show impressions of tentacles emerging from the end. *Conularia* is often found in muddy sediments, and having the ability to close the entrance of the shell might have prevented the animal being choked by fine sediment. Another interpretation suggests that the animal floated in the sea pointed end upwards, with the animal hanging below.

This is an example of an extinct animal group that had a long history, but without knowledge of the soft parts of the animal we can only speculate on the type of animal that inhabited the shell. It may have looked like a sea anemone or jellyfish in a cup.

21 *Conularia.*

WORMS

Worms are abundant in the world today, and have existed since the Precambrian. They are found in a great range of environments, and are an exceptionally important component of both soil and sea floor faunas. The vast majority are soft-bodied and therefore rare as fossils. However, many types are active burrowers and leave distinctive trace fossils, which are common in the geological record. A few types of worms are common because they build a solid tube in which they live, such as the white calcareous tubes of serpulid worms that are common on rocky shores today. Some polychaete worms, like the modern ragworm, have strong jaws that are found as fossils.

Rarely, the impressions of soft-bodied worms are preserved in rocks, and exceptionally, three-dimensional preservation is known in amber, and in the example of a nematode worm from the Rhynie chert illustrated below.

22

Palaeonema / Nematode Worm / Devonian

Phylum	Nematoda
Class	Enoplia
Gen et sp	***Palaeonema phyticum***
Locality	Rhynie, Aberdeenshire
Age	Early Devonian
Stratigraphy	Rhynie Cherts Member, Dryden Flags Fm., Old Red Sandstone

Nematode worms are exceptionally common in virtually all environments today, but since they are small and soft-bodied they have virtually no fossil record. Until recently the oldest undoubted fossil nematodes were of Early Cretaceous age from Lebanon. Most fossil examples of nematodes have been found in amber. The finding of many tiny fossil nematodes, with adults, juveniles and eggs within

22a Cross-section of *Aglaophyton* axis, containing the nematode *Palaeonema* (arrowed).

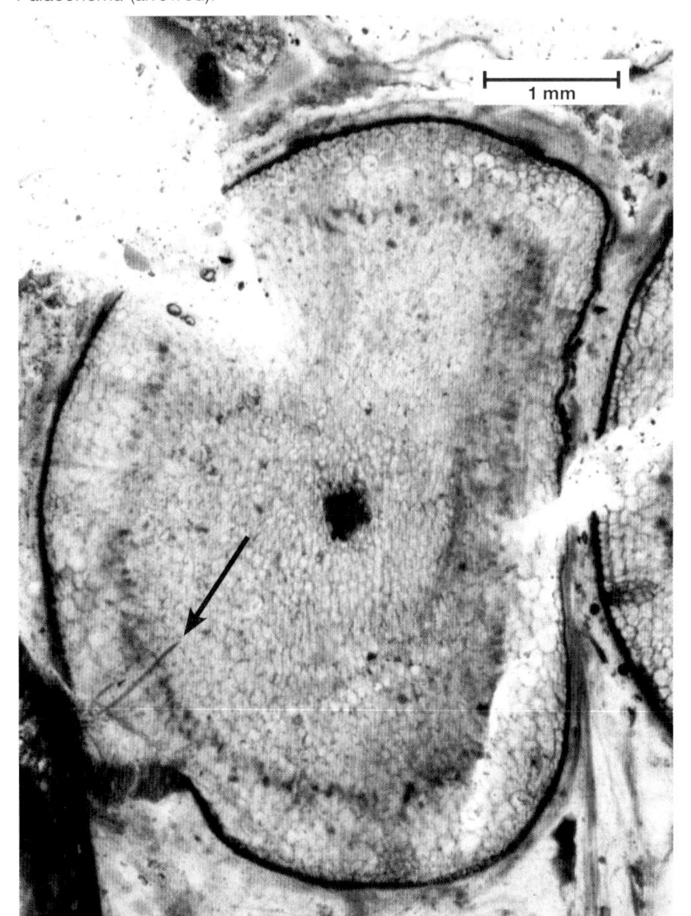

22b The oldest known nematode worm, *Palaeonema* in an axis of *Aglaophyton*.

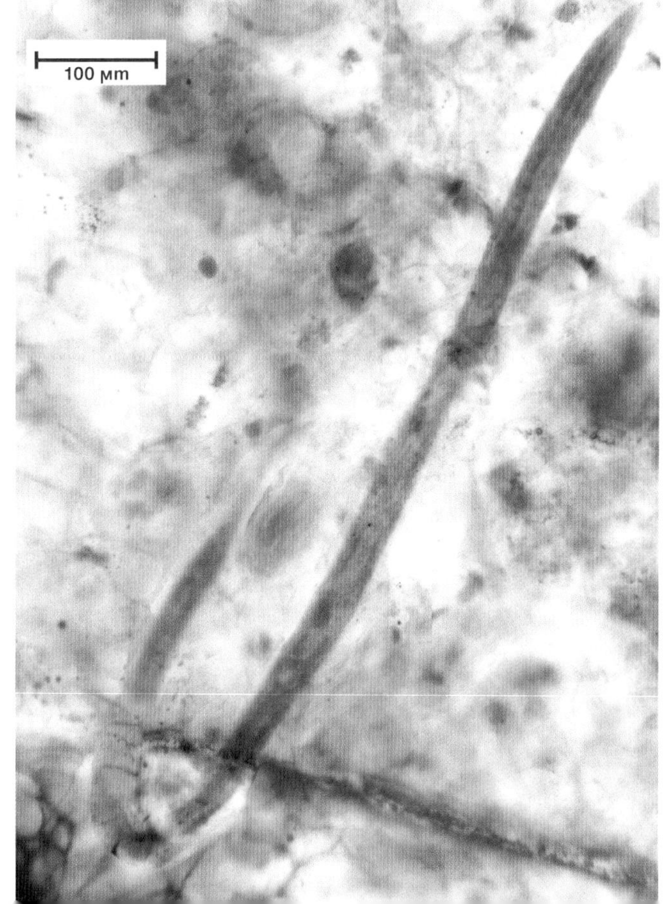

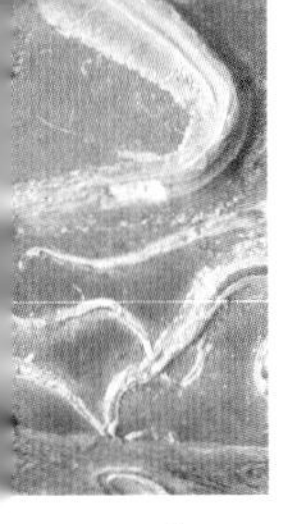

partially decayed plant stems of *Aglaophyton* in the Rhynie chert takes the fossil record of this group back to the Early Devonian, over 400 million years ago. This exceptional fossil shows that nematodes invaded land plants very early in the history of plant development (Poinar *et al.* 2008). This is the oldest evidence of an association between a plant and an animal. The nematodes were probably parasites, and may have played a role in the decomposition of vegetation, just as they still do today. At the present day nematodes invade plants by entering through the pores (stomata) in the plant stem that the plant uses to take in carbon dioxide, and then eat cell material within the plant. *Palaeonema* probably entered the *Aglaophyton* stem by the same route. The exceptional preservation in the Rhynie chert (*see* plants section) allows this glimpse into the past, and also illustrates the poor nature of the fossil record of soft-bodied animals.

22c Detail of tip of *Palaeonema*.

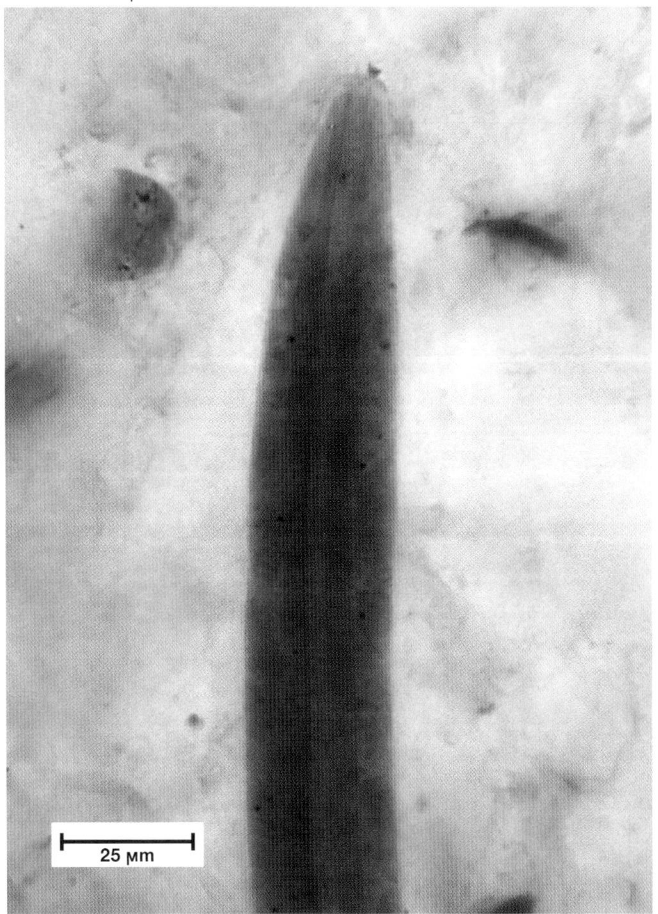

23

Serpula / Tube Worm / Jurassic

Phylum	**Annelida**
Class	**Polychaeta**
Gen et sp	***Serpula***
Locality	**Dunans, Staffin, Isle of Skye**
Age	**Late Jurassic, Oxfordian**
Stratigraphy	**Staffin Shale Formation**

This is a belemnite guard that was colonised by serpulid worms while it lay on the sea floor, and before it was buried. The white calcareous tubes are concentrated on one side of the specimen. The worms used the belemnite as a point of attachment on the muddy sea floor and were able to filter food particles from the water. Serpulid worms are a successful and long-lived group with a world-wide distribution, and modern examples can be found encrusting shells washed up on many beaches in Scotland.

23 Calcareous tubes of the worm *Serpula* attached to a belemnite.

BRYOZOANS

Bryozoans are very common colonial animals in modern seas and have a geological record extending back to the Ordovician. There are over 3000 modern species and over 15,000 fossil species and yet most people have never heard of them. They have a low profile because they are generally small; colonies seldom exceed a few centimetres in size, and a hand lens is needed to see the individual animals and the detail of the morphology. They secrete a calcareous or organic skeleton of chambers in which individuals live; each has tentacles to gather food, a mouth, gut and anus. Bryozoan colonies are often cemented to other objects, and are commonly found attached to marine fossils such as sea urchins, brachiopods and molluscs. Some types form colonies that grow into sheets, nets or stick forms.

On British seashores they can be found cast up on the strandline, and attached to rocks and seaweed on the lower shore.

24a *Fenestella* colony in shale with parts of a trilobite at lower right.

24

Fenestella / Bryozoan / Carboniferous

Phylum	**Bryozoa (Ectoprocta)**
Class	**Stenolaemata, Fenestrata**
Gen et sp	***Fenestella***
Locality	**Bishop Hill, Kinross**
Age	**Early Carboniferous**
Stratigraphy	**Charlestown Main Limestone, Lower Limestone Formation**

Fenestella colonies were abundant in the warm sea of the Carboniferous. They grew as sheet-like forms forming a loose inverted cone, or curved fan shape. Colonies were rooted in sediment and grew to about 10 cm in height off the sea floor. They were frequently crowded together, and must have formed bryozoan carpets over areas of sea floor. The colonial sheet is made of branches with regular connecting bars between the branches. The animals lived in the branches, and the apertures through which the animals extended their tentacles are all on the outside of the sheet structure. It is thought that when feeding the animals trapped food particles from currents passing through the holes in the structure. The apertures in the branches are so placed that each animal had access to part of one of the holes or 'fenestrae' in the structure.

The fan-like structures also had the effect of trapping sediment, so they are sometimes found preserved in growth position, but they are more frequently seen flattened on bedding surfaces, as in the example shown. The colony illustrated also shows growths used to stabilise the colony.

24b Detail of *Fenestella* colony showing network of branches.

BRACHIOPODS

Brachiopods, like bivalved molluscs, have a shell with two valves, but here the similarity ends. The two valves are different sizes and are symmetrical about a median plane; these valves are dorsal and ventral to the animal, which posesses a structure called a lophophore for filtering food particles from sea water. Most brachiopods have a short flexible stalk or pedicle that emerges from an opening in the ventral valve, and is used to fix the animal to objects on the sea floor. Other brachiopods were cemented to the substrate, or free living.

Three classes of brachiopods are recognised. Two are termed 'inarticulate' because they lack teeth and sockets to articulate the valves. One class, the Lingulata, have shells of chitinophosphatic composition, and the other, Inarticulata, have calcareous shells. (Until recently both these classes were grouped together in the Inarticulata). The third class, the Articulata, have calcareous shells with teeth and sockets to hinge the valves.

Brachiopods were the dominant shellfish of the Palaeozoic, and lived in marine environments.

Much brachiopod research has concerned the 'community ecology' of these animals, so we can trace the brachiopod groups through time and determine the environments they favoured. In general, strong shells with ribs favoured shallow water and coarse sediment types, whilst thinner-shelled, smooth forms lived in deeper water with muddier sediment. Brachiopods declined rapidly at the end of the Palaeozoic, and the molluscs expanded rapidly to become the dominant shellfish in the sea. Thus former brachiopod habitats rapidly became dominated by bivalves. The bivalves, with siphons for feeding, could remain hidden in the sediment and were possibly less vulnerable to predation by fish, and newly evolved predatory starfish. However, this does not explain the continuation to the present day of two orders typified by *Terebratula* and *Rhynchonella*. Maybe they are just not so tasty, and not abundant enough for a predator to specialise in eating them. Many brachiopods still exist today, even around Scottish coasts, but they are generally found in deeper offshore areas and their shells are rarely found on beaches.

25

Lingula / Brachiopod / Carboniferous

Phylum	**Brachiopoda**
Class	**Lingulata**
Gen et sp	***Lingula***
Locality	**Seafield, Fife**
Age	**Early Carboniferous**
Stratigraphy	**Lower Limestone Formation, below 2nd Abden Limestone**

Lingula is a survivor! Shells that are virtually indistinguishable from modern forms are found in rocks as far back as the Cambrian. Today, *Lingula* lives in mud in estuaries and deltas. Throughout the geological record *Lingula* is associated with muddy conditions at the marine margin, and seems to have been able to thrive in brackish conditions. In Fife the Second Abden Limestone is marine in origin, and the shales below the limestone probably represent the initial part of a marine transgression when a muddy delta was being covered by the sea, thus fully marine conditions had not been fully established.

The animal is a filter-feeder and lives in a vertical burrow where it is fixed by a long, flexible pedicle. The pedicle is a gastronomic delicacy in parts of the East, but is unlikely to become popular in Scotland, even deep-fried.

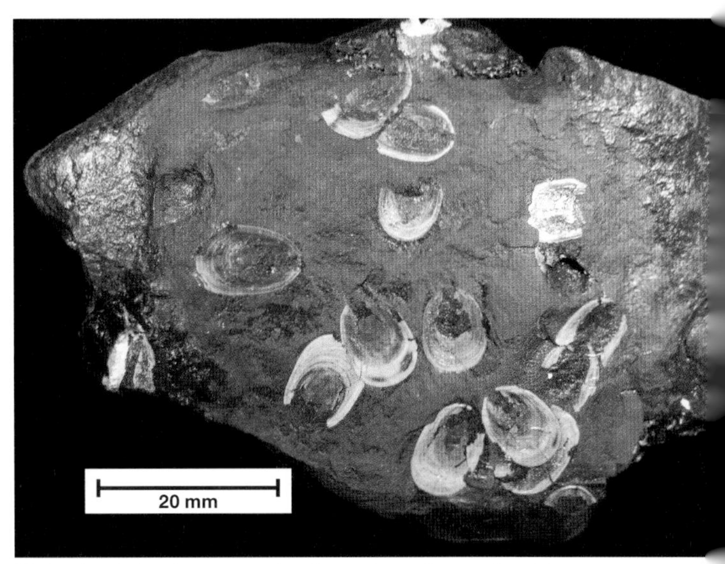

25 Shells of *Lingula* preserved on the surface of a small nodule.

26

Leptaena / Brachiopod / Silurian

Phylum	**Brachiopoda**
Class	**Articulata**
Gen et sp	***Leptaena***
Locality	**South Threave, Girvan, Ayrshire**
Age	**Ordovician**
Stratigraphy	**Starfish Bed, South Threave Fm., Drummock Group**

This common brachiopod of the Ordovician and Silurian periods has a distinctive shape and ornament. The shape is termed 'geniculate' since the shell has a knee-shaped bend which served to raise the shell openings above the mud of the sea floor. The ornament consists of fine radial ribs that are crossed by irregular concentric ridges. The shell was made of calcite, but specimens in the Starfish Bed have usually had the shell dissolved away to leave an external or internal mould of the shell. The illustration shows an external mould displaying the shell ornament, together with an isolated internal mould showing the inside surface of the shell with scars of the muscles that held the shell closed in life. The two look very different, and this is frequently the case when internal and external moulds of fossils are found.

26 *Leptaena* preserved as an external impression (in rock), and as an internal cast.

27

Schuchertella / Brachiopod / Carboniferous

Phylum	**Brachiopoda**
Class	**Articulata**
Order	**Strophomenida**
Gen et sp	***Schuchertella***
Locality	**Bishop Hill, Kinross**
Age	**Early Carboniferous**
Stratigraphy	**Lower Limestone Formation, Charleston Main Limestone**

These brachiopods occur with other brachiopods, abundant bryozoans, crinoids, and occasional trilobites and cephalopods in Carboniferous shaly limestones on Bishop Hill. They can be found on the tips from old limestone workings. In this rich sea-floor community the adult *Shuchertella* appear to have lacked a functional pedicle, and lay on the sediment surface. They lived by filtering food particles from the sea water. Many of the specimens that come from the south end of the hill beneath a dolerite sill have been altered to rusty moulds, and the carbonate shell material has dissolved. The shales are also hard and flinty, having been baked by heat from the sill that was intruded into the sediment. The hard dolerite is responsible for the scarp formation of Bishop Hill.

27 Rusty impressions of *Schuchertella* in baked shale.

28

Gigantoproductus / Brachiopod / Carboniferous

Phylum	**Brachiopoda**
Class	**Articulata**
Order	**Strophomenida**
Gen et sp	***Gigantoproductus giganteus***
Locality	**Inverteil, Fife**
Age	**Early Carboniferous**
Stratigraphy	**Lower Limestone Formation**

This large brachiopod had a thick shell and lay on the sea floor on its convex valve. The thick shell helped to keep it in a stable position so that the margins of the shell were clear of the sediment surface and it could filter feed by passing water through the lophophore which trapped food particles. The shell was concavo-convex, and this left little room for the body of the animal between the valves of the shell. In overall shape and mode of life it is similar to the oysters (see *Gryphaea*) of the Mesozoic, which appear to have taken over a similar environment following the extinction of the Productid brachiopods in Permian times.

28 *Gigantoproductus*, one of the largest brachiopods.

29

Spirifer / Brachiopod / Carboniferous

Phylum	**Brachiopoda**
Class	**Articulata**
Order	**Spiriferida**
Gen et sp	***Spirifer***
Locality	**Roscobie, Fife**
Age	**Early Carboniferous**
Stratigraphy	**Lower Limestone Formation**

Spirifer has a characteristic shape with a wide hinge line, and ribbed valves. There is a prominent fold in the dorsal valve which matches a groove in the ventral valve. Inside the shell the lophophore was supported by a spiral structure made of calcite. Water was drawn into the shell at the shell margins, and passed through the lophophore to extract food, with the exhalent current emerging from the fold in the shell margin. Spiriferid brachiopods were diverse and abundant in the seas during the Devonian and Carboniferous periods, but only a few survived into the Jurassic, and they were extinct before the start of the Cretaceous.

29 Specimens of *Spirifer* extracted from shale.

MOLLUSCS

Gastropods, bivalves and cephalopods are the main groups of molluscs found as fossils. Many molluscs have hard calcareous shells that are easily fossilised. Thus there is a good fossil record of bivalves, gastropods and ammonites, but there is a poor fossil record for those, like the slugs, that lack significant hard parts. Most molluscs live in the sea, but some bivalves and gastropods live in fresh water, and gastropods have successfully invaded the land to give the great variety of land snails in the world today.

The examples chosen are typical Scottish representatives of the main molluscan groups found as fossils.

Molluscs - Cephalopods

Cephalopods are the group of molluscs that include modern squid, cuttlefish, octopus and the pearly nautilus. Of these modern animals, only the nautilus has an external shell. In the past there were many more shelled cephalopods, notably the ammonoids, of which ammonites are commonly found in the Jurassic rocks of Scotland. The animal was similar to a squid in a shell; it had tentacles, good eyes, and a type of jet-propulsion mechanism operated by drawing water into an internal mantle cavity, and then squirting it out through a narrow tube (hyponome). The shell was chambered, and the animal could control its buoyancy by adjusting the quantities of gas or fluid in the chambers. The line where the chamber wall joins the outer shell is known as the suture line; this line is a simple curve in *Nautilus*, zig-zag in goniatites, but complex in ammonites.

All cephalopods lived in marine conditions, and the goniatites and ammonites evolved very rapidly, making them ideal as 'Zone Fossils', characterising short periods of geological time. Ammonite faunas in the Jurassic are known in such detail that many of the ammonite 'Zones' lasted less than half a million years.

There are five types of fossil cephalopods that are found in Scotland. These are:

- Straight-chambered shells of orthocones (Cambrian to Carboniferous)
- Coiled *Nautilus* shells with simple chambers (mainly Jurassic)
- Goniatites, with zig-zag chamber margins – Carboniferous in Scotland and rare
- Ammonites that have complex frilled chamber margins (mainly Jurassic)
- Belemnites, which were internal to a squid-like animal, and are analagous to part of a modern cuttlefish bone (mainly Jurassic).

30

Cycloceras / Orthocone Nautiloid / Carboniferous

Phylum	Mollusca
Class	Cephalopoda
Order	Orthoceratida
Gen et sp	*Cycloceras*
Locality	Bishop Hill, Kinross
Age	Carboniferous
Stratigraphy	Lower Limestone Formation

Orthocone cephalopods such as *Cycloceras* had straight, tapering chambered shells, secreted by the animal as it grew. Most examples are small, but a few types had shells a metre long. The animal probably crawled along the sea bottom using its tentacles, but would also have been capable of swimming. Some of these animals had the ability to balance the shell in a horizontal position for swimming by depositing calcite in the chambers to act as a balance weight. When swimming using jet-propulsion it would have been swimming backwards, raising the question of how it could avoid collisions. However, modern squid swim backwards and have the eyes on the side of the head, so they probably had virtually all-round vision. The example shown has annular corrugations of the shell; others were longitudinally ribbed, but the majority had smooth shells. This specimen is from the Charleston Main Limestone and it probably lived in association with the many crinoids, brachiopods and bivalves that occur at the same locality. A speculative interpretation might be that the ornament of *Cycloceras* acted as camouflage amongst the large segmented crinoid stems, and it stalked its prey with the shell held vertically amongst the crinoids.

30 The orthocone *Cycloceras* crushed in shale.

31

Cravenoceras / Goniatite / Carboniferous

Phylum	Mollusca
Class	Cephalopoda
Order	Goniatitida
Gen et sp	*Cravenoceras scoticum*
Locality	East Kilbride
Age	Carboniferous, Namurian
Stratigraphy	Upper Limestone Formation, shales above Calderwood Cement Lst

Goniatites are small cephalopods that evolved rapidly in Devonian and Carboniferous times. They are exclusively marine in origin, and were free-swimming in the oceans. Thus they are very useful for correlation of marine Devonian and Carboniferous rocks in the south of Britain. *Cravenoceras* is one of the genera useful for zonation of the Namurian stage of the Carboniferous. In the Namurian basins of England goniatites are found in 'Marine Bands' within deep-water shales, and occur with thin-shelled bivalves (*Posidonia*), nautiloid cephalopods, rare fish, conodonts and radiolaria. In Scotland rocks of similar age are mostly non-marine or at best near-shore shallow marine in origin, and goniatites are very scarce. Thus it is difficult to correlate the rocks between Scotland and England using goniatites, but the English succession can be correlated over continental Europe and Ireland, and even to America.

31 Partly crushed specimens of the goniatite *Cravenoceras.*

32

Coroniceras / Ammonite / Jurassic

Phylum	Mollusca
Class	Cephalopoda
Order	Ammonitida
Gen et sp	*Coroniceras*
Locality	Broadford, Skye
Age	Early Jurassic
Stratigraphy	Pabay Shale Formation (= Upper Broadford Beds)

This is a typical large Early Jurassic ammonite with a lot of whorls exposed and simple ribs on the shell. This one is from the Upper Broadford Beds (now included in the Pabay Shale Formation of Hesselbo *et al.* 1998), and examples occur on bedding surfaces on the beach in the Broadford area. They are impressive in size, but poorly preserved, so detailed identification to species level is frequently impossible when diagnostic features have been lost.

The animal lived in part of the outer whorl of the shell, and the rest was divided into chambers that acted as a buoyancy mechanism. By controlling the amount of water or gas in the chambers, the animal could rise or sink in the water rather like a submarine.

This ammonite had tentacles for gathering food, and it probably had a varied diet of molluscs, arthropods and fish similar to modern octopus and squid. It might have been preyed on by the marine reptiles of the time such as *Ichthyosaurus* and *Plesiosaurus.*

32 Large specimen of the Lower Jurassic ammonite *Coroniceras.*

33

Dactylioceras / Ammonite / Jurassic

Phylum	**Mollusca**
Class	**Cephalopoda**
Order	**Ammonitida**
Gen et sp	***Dactylioceras***
Locality	**near Holm Island, Bearreraig Bay, Trotternish, Isle of Skye**
Age	**Early Jurassic, Toarcian**
Stratigraphy	**Raasay Ironstone Formation**

Dactylioceras occurs in shales at the base of the Raasay Ironstone near the disused iron mine on Raasay at a point where the ironstone has been quarried at the surface, leaving the underlying shales exposed on a dipping bedding plane. On areas that are not overgrown, near the small cliff of ironstone there are many flattened *Dactylioceras*, and also numerous belemnites. The illustrated specimens are preserved uncrushed in a nodule, and are probably a different species from those seen at the Raasay mine. Rapid evolution of these forms produced several species of *Dactylioceras* and closely related genera.

This fossil is more famously found at Whitby in Yorkshire, where the specimens occur uncrushed in nodules. In the past people in Yorkshire carved heads on the end of the coil and sold them to gullible believers as snakes petrified by St Hilda.

33b *Dactylioceras* extracted from matrix.

33a Small *Dactylioceras* crowded in a nodule.

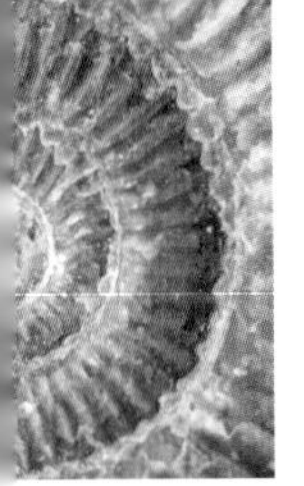

34

Ludwigia / Ammonite / Jurassic

Phylum	**Mollusca**
Class	**Cephalopoda**
Order	**Ammonitida**
Gen et sp	***Ludwigia murchisonae*** **(the larger specimen)**
Locality	**Bearreraig Bay, Trotternish, Isle of Skye**
Age	**Mid-Jurassic, Bajocian**
Stratigraphy	**Bearreraig Sandstone Formation**

The cliffs at Bearreraig Bay to the north of Portree on Skye are capped by a dolerite sill of Tertiary age, below which is the Bearreraig Sandstone Formation, which is of Middle Jurassic age and marine origin. The sandstones are fossiliferous and ammonites occur at several levels, as described by Morton (1965). Ammonites of the genus *Ludwigia* are common in calcite-cemented concretions that occur within four beds of sandstone at the base of the Bearreraig Sandstone Formation on the shore at the south end of Bearreraig Bay. For each species of ammonite there are two forms present; one is large (c.20cm diameter) and has a smooth outer whorl of the shell. The other is about 7cm in diameter, having a ribbed shell with projections called lappets at the mouth of the shell. The large form is interpreted as a female and the small form as a male. Many modern cephalopods (e.g. squid) also show sexual dimorphism with the female much larger than the male.

Several related genera are present at this locality, but *Ludwigia murchisonae* has been chosen as a zone fossil for part of the Bajocian stage of the Jurassic period. Ammonites evolved rapidly and are ideal for correlating rocks from different areas. In Britain the sandstones at Bearreraig can be correlated with rocks on the coast at Bridport in Dorset on the basis of the ammonites they contain. The specific name *murchisonae* is in honour of Lady Murchison, wife of the famous geologist Sir Roderick Impey Murchison who wrote *The Silurian System*, one of the classics of nineteenth century geology.

Other fossils present at this locality are bivalves, belemnites and fossil wood. The ammonites lie at random angles in the rock, rather than lying flat, as might be expected. It is probable that the shells were moved by the activities of burrowing animals that disturbed (bioturbated) the sediment soon after the shells were buried on the sea floor. At this locality it is not permitted to damage the outcrop by attempting to extract fossils; loose material can generally be found on the beach.

34 Female (large) and male dimorphs of the ammonite *Ludwigia*.

35

Pictonia / Ammonite / Jurassic

Phylum	**Mollusca**
Class	**Cephalopoda**
Order	**Ammonitida**
Gen et sp	***Pictonia* cf *involuta***
Locality	**Shandwick, Ross-shire**
Age	**Late Jurassic, Kimmeridgian**
Stratigraphy	**Kimmeridge Clay Formation**

This fine ammonite is from the Hugh Miller Collection in Royal Museums, Scotland, and was probably collected by Hugh Miller on one of his collecting excursions from his home in Cromarty to the Jurassic sections on the shore at Shandwick. Most ammonites in the shaly Kimmeridgian rocks of the area are crushed flat, but this one was preserved in a nodule and retains its three-dimensional shape. The outer shell has been lost, and the specimen clearly shows the complex suture lines that represent the edges of the chamber walls where they met the outer shell. Thus we can see the chambered part of the shell, and the body chamber in which the animal lived. The hollow chambers have been filled with calcite, but sediment fills the body chamber. Since the outer shell is missing the details of the external ornament of the shell cannot be seen. The name given here is as given on the museum label, but probably requires revision.

35 The ammonite *Pictonia* from the Hugh Miller Collection, National Museums Scotland.

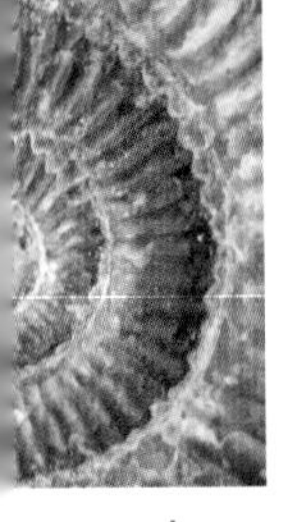

36

Cylindroteuthis / Belemnite / Jurassic

Phylum	**Mollusca**
Class	**Cephalopoda**
Order	**Belemnitida**
Gen et sp	***Cylindroteuthis ?puzosiana***
Locality	**Brora foreshore, Sutherland**
Age	**Jurassic, Oxfordian**
Stratigraphy	**Brora Brick Clay, Brora Argillaceous Formation**

The bullet-shaped fossil belemnite is part of the internal structure of a squid-like animal, and is partly analagous to the 'bone' of the modern cuttlefish. The belemnite 'guard' was situated in the tail of the animal. The conical cavity at the blunt end held a fragile chambered shell, the phragmocone, that is seldom preserved. *Cylindroteuthis*, along with other belemnite species, can be found in the Brora Brick Clay on the shore south of the mouth of the Brora River at low tide, providing the rocks are not covered with beach sand. The belemnites are usually fractured into several sections and have to be carefully excavated and then washed, dried and glued back together. Fragments of belemnites also occur loose on the beach, but be careful – there was a wartime shooting range near the locality, and bullets looking like belemnites may also be found!

Very rare examples of belemnites have been found elsewhere with impressions of the soft body and tentacles preserved. Belemnites posessed an ink sac, and Victorian collectors wrote letters in reconstituted belemnite ink. Belemnites were food for some of the marine reptiles of the Jurassic, and their remains occur as stomach contents in fossil reptiles.

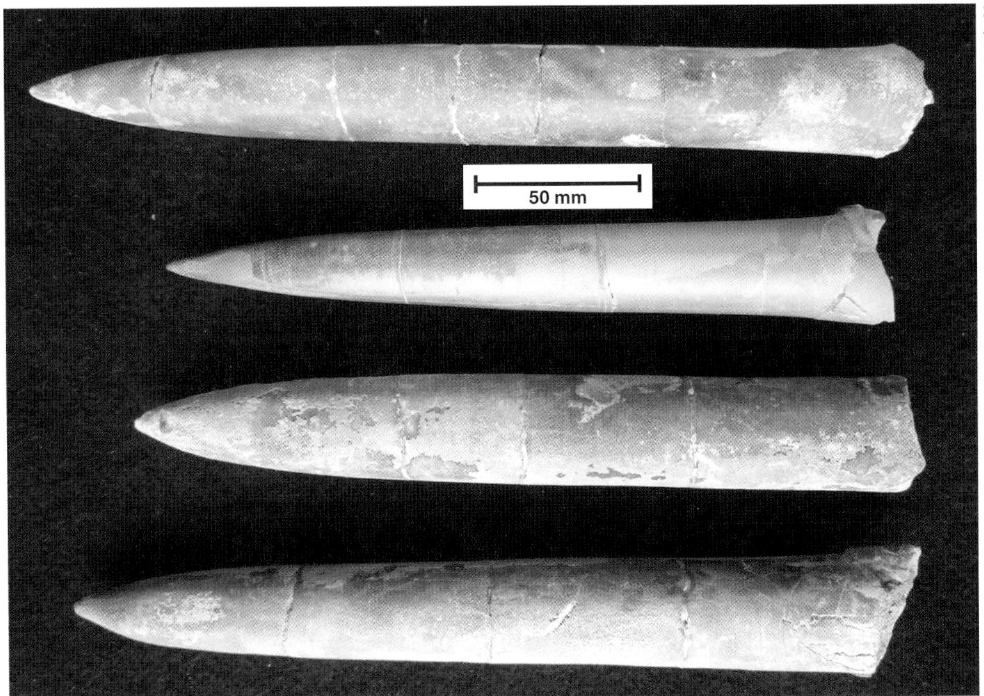

36a Belemnite guards, *Cylindroteuthis*, from shales at Brora.

36b Reconstruction of a belemnite, cut away to show internal guard and phragmocone.

37

Megateuthis / Belemnite / Jurassic

Phylum	**Mollusca**
Class	**Cephalopoda**
Order	**Belemnitida**
Gen et sp	***Megateuthis* sp.**
Locality	**Bearreraig Bay, Skye**
Age	**Jurassic, Bajocian**
Stratigraphy	**Bearreraig Sandstone Formation**

As the name implies, *Megateuthis* is a large belemnite. Several species have been described, and some of the largest belonged to an animal probably about 2m long. Such a creature would have been a significant predator in the Jurassic seas around Skye. The guard of *Megateuthis* has an oval cross-section, and the tip is flattened laterally and has distinct grooves on the flanks. The chambered part of the shell, known as the phragmocone, is frequently preserved within the cone-shaped cavity at the front (thick) end of the guard.

37 *Megateuthis*, parts of the phragmocone and guard.

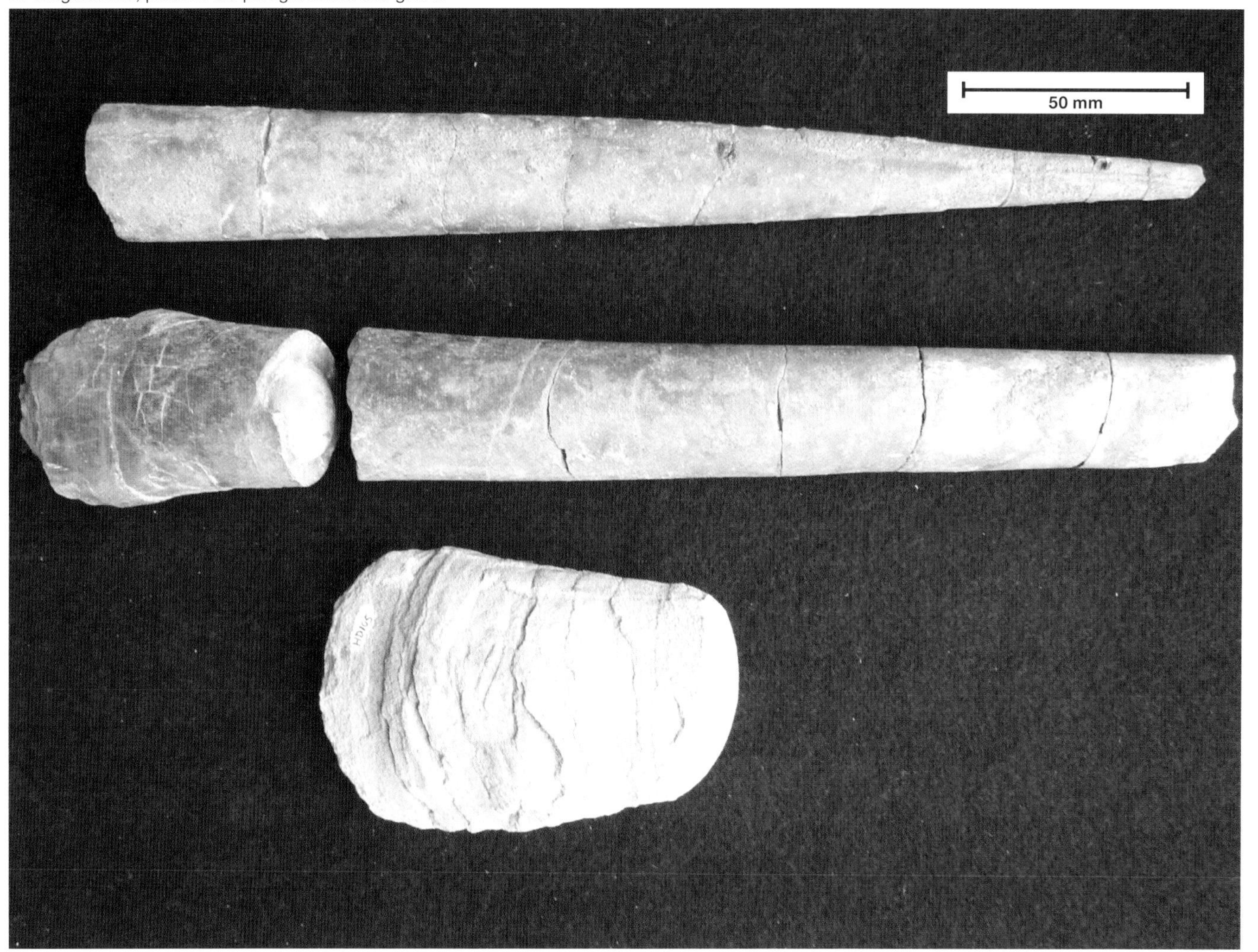

Molluscs - Bivalves

Bivalve molluscs have a shell that comprises two valves. The valves are placed either side of the animal, and are hinged together. A ligament serves to open the valves when the animal is feeding, and internal muscles are contracted to keep the valves closed. In most bivalves the left and right valves are mirror images of each other (e.g. cockles), but in some bivalve groups, particularly the oysters, one valve is much larger than the other. Some bivalves, such as oysters and mussels, live on the surface, and are cemented to rocks, or attached to objects by organic threads. Others, such as scallops, are free living on the seabed and can swim by flapping the valves. Many others burrow in the sediment, feeding by means of siphon tubes that are extended from the burrow. A few bivalves are capable of boring into rock, and leave distinctive holes that can be preserved as a trace fossil.

38

Poldevicia / Bivalve / Carboniferous

Phylum	**Mollusca**
Class	**Bivalvia**
Gen et sp	***Polidevcia attenuata***
Locality	**Roscobie, Fife**
Age	**Early Carboniferous**
Stratigraphy	**Lower Limestone Formation**

This elegant little bivalve occurs in shales and limestones along with crinoids, brachiopods and other bivalves. The elongated posterior end of the shell indicates that it probably burrowed in the sediment, and only the thin tip of the shell would be exposed on the surface when it was feeding. Thus it was probably well camouflaged on the sea floor. The simple ornament of shallow concentric grooves is also typical of burrowing types of bivalve. The main predators of shellfish in the Carboniferous were probably cephalopods, and fish. There were a number of types of early shark that had flat teeth suitable for crushing shells, and they probably dug in the sediment for food such as bivalves, shrimps and worms in the same way as modern rays hunt for food.

38 *Poldevicia*, small bivalves that burrowed in a muddy sea floor.

39

Gryphaea / Bivalve / Jurassic

Phylum	**Mollusca**
Class	**Bivalvia**
Gen et sp	***Gryphaea arcuata***
Locality	**Waterloo, Broadford, Isle of Skye**
Age	**Early Jurassic**
Stratigraphy	**Pabay Shale Formation (= Upper Broadford Beds)**

This is one of the few fossils to have a common name, being known as 'devil's toenails' in some parts of Britain. *Gryphaea* is in fact an oyster that lived in muddy sand in shallow seas of the Early Jurassic. It is common on the shore at Waterloo near Broadford, where one bed is full of these fossils. The two valves of the shell are different in shape, as is common in oysters; one forms an open coil, and the other is flat (the 'toenail') and forms a lid. The shell is thick, so it was a strong shell and easily fossilised. Where conditions were suitable this oyster virtually covered the seabed. A few other fossils are present at the locality, including scarce ammonites and small bivalves similar to modern scallops.

Gryphaea occurs at other localities around Broadford, and in rocks of the same age on the shores of Loch Slapin about 1km south of Camas Malag, from where there are magnificent views of Bla Bheinn. The stratigraphic nomenclature of the Jurassic rocks on Skye has been modified by Hesselbo *et al.* (1998). The Upper Broadford Beds and Pabba Shales of previous schemes have now been combined as the Pabay Shale Formation.

39 *Gryphaea*, an oyster known as 'Devil's toenails'.

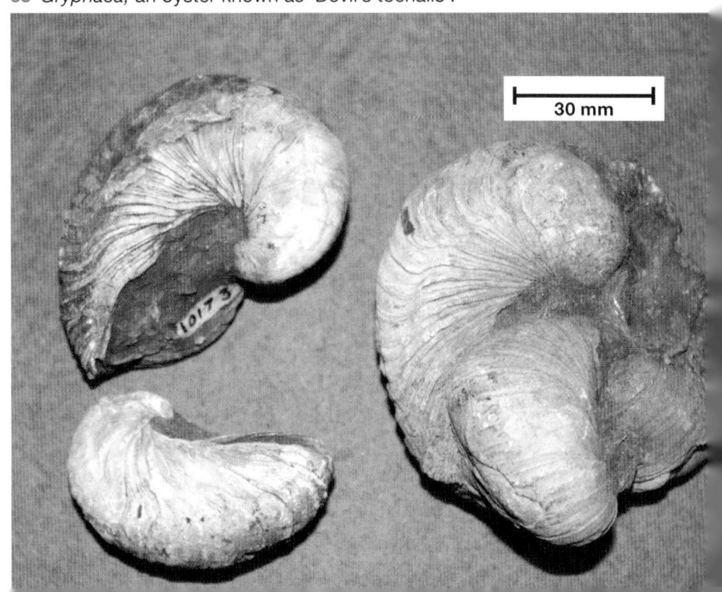

40

Hippopodium / Bivalve / Jurassic

Phylum	**Mollusca**
Class	**Bivalvia**
Gen et sp	***Hippopodium ponderosum***
Locality	**Golspie, Sutherland**
Age	**Early Jurassic**
Stratigraphy	**Lady's Walk Shales, Dunrobin Bay Formation**

This is a characteristic bivalve of the Lower Lias in England, and can be found in Scotland in the Pabba Shales on Raasay, and in the small shore exposures of the Lady's Walk Shales on the shore near Dunrobin Castle at Golspie, Sutherland. Its name translates as 'ponderous horse foot', and it certainly has a passing resemblance to a hoof. The shell is very thick and strong, and the umbones are twisted so that each valve is an open spiral. The shell lay on the muddy sea floor with the plane between the valves vertical. It was a filter feeder, and thus had to avoid being covered with mud. Powerful muscles kept the valves shut tightly for protection, but relaxed to allow the valves to open for feeding. The valves were opened by the ligament, which acted like a flexible external spring. This specimen is unusual in that the chitinous ligament is preserved, and it is very thick. Presumably it needed to be large and powerful to open the heavy valves, which were lying on sticky mud.

40a and b Views of *Hippopodium*, showing the twisted umbones and coarse ornament.

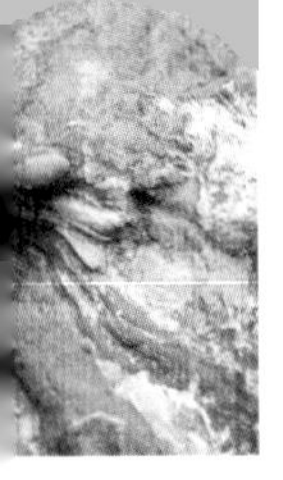

41

Pseudopecten / Bivalve / Jurassic

Phylum	Mollusca
Class	Bivalvia
Gen et sp	*Pseudopecten aequivalvis*
Locality	Rubh na Leac, Isle of Raasay
Age	Jurassic, Middle Lias, Domerian
Stratigraphy	Scalpa Sandstone Formation

The Scalpa Sandstone forms the high cliff to the south of Rubh na Leac, and is separated by a fault from the red Triassic sandstones and conglomerates exposed on the point. In the tumble of fallen sandstone blocks on the beach there are some surfaces that are littered with large fossil scallop shells, looking remarkably similar to modern scallops that are fished in the area today. These Jurassic scallops lived on a sandy sea floor with oysters, a few brachiopods and crinoids. The sandstones are bioturbated (churned by burrowing animals) indicating a rich fauna that lived both in and on the sandy sea floor.

41 *Pseudopecten* preserved mainly as impressions in sandstone.

42

Goniomya / Bivalve / Jurassic

Phylum	Mollusca
Class	Bivalvia
Gen et sp	*Goniomya literata*
Locality	Clynelish Quarry, Brora, Sutherland
Age	Late Jurassic
Stratigraphy	Clynelish Quarry Sandstone, Brora Arenaceous Formation

Goniomya is a bivalve that burrowed into sands in a shallow sea bordering the Jurassic Scottish island in the late Jurassic. It has a highly characteristic ornament of ridges that form a repeated v-shape, and is also known as *Goniomya v-scripta*. The Clynelish Quarry sandstone contains a varied fauna of bivalves that favoured a sandy nearshore environment, together with scarce ammonites and numerous casts of fossil wood fragments. One of the two valves was broken prior to burial, and this may have been due to predation, the most likely attackers being fish, crabs or cephalopods.

Clynelish Quarry was the source of much of the building stone of Brora, but working ceased many years ago and it was filled with refuse. Only a small exposure remains, and the quarry area is now an industrial site. However, fossils can still be seen in buildings in the area, notably an ammonite in the war memorial in Brora, and numerous specimens in the large blocks used to build parts of Dunrobin Castle at Golspie. There are also some fossils from the Clynelish Sandstone on show amongst the eclectic collection of artifacts in Dunrobin Castle Museum, near Golspie. Many of these specimens were presented to the museum by John W. Judd, who had worked with the honorary museum curator, the Revd Dr Joass, on the Jurassic of the Brora area (Judd 1873).

42 *Goniomya*, a burrowing bivalve with distinctive ornament.

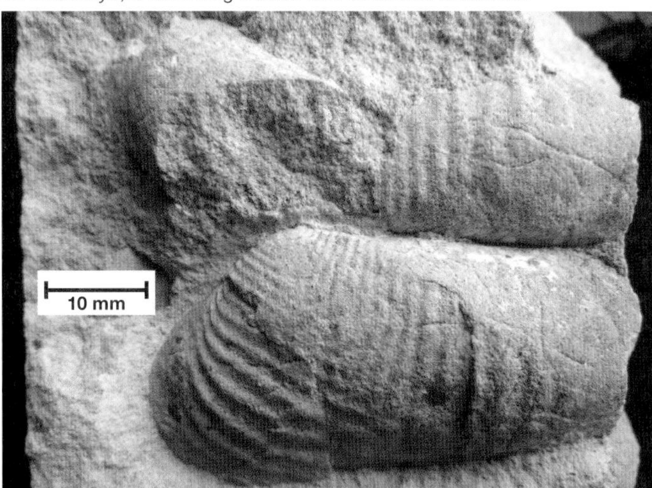

43

Lithophaga / Bivalve / Jurassic

Phylum	**Mollusca**
Class	**Bivalvia**
Gen et sp	***Lithophaga***
Locality	**Helmsdale, Sutherland**
Age	**Late Jurassic**
Stratigraphy	**Helmsdale Boulder Beds, Kimmeridge Clay Formation**

Lithophaga bores into solid rock! Its name means rock-eater, but of course the animal does not eat rock, but makes the bored hole for protection, and safe in its home it filters water for food particles. On the beach at Helmsdale it is possible to find both modern and fossil rock-boring bivalves. Rocks thrown up on the beach by storms have holes that sometimes still contain the shell of the modern rock borer. More uncommon are rocks within the Boulder Beds that have fossil borings, now with a sediment fill that has been turned to rock. A few of these borings still contain the fossil shell responsible for making the boring.

The bivalves employ a combination of mechanical and chemical action to assist boring. They start to make the hole when they are small, but as they bore deeper, and the shell grows, so the boring expands into a vase shape, and the shell becomes too big to get out of the entrance. If two bivalve borings meet, then both animals will die, since the protective homes of both have been broken. Most of the fossil borings made by *Lithophaga* are in calcareous flagstone boulders from the Helmsdale Boulder Beds, or in colonies of the coral *Isastraea*.

43a *Lithophaga* shells preserved in their borings in a pebble from the Helmsdale Boulder Beds.

43b A boulder in the Helmsdale Boulder Beds with numerous *Lithophaga* borings.

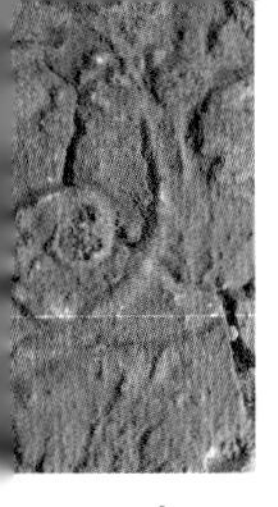

Molluscs - Gastropods

Gastropods are molluscs that have a shell that is generally twisted into a spiral shape, and a muscular foot that it uses to crawl about and to cling to surfaces. Garden slugs and snails, as well as whelks and limpets from the seashore, are all gastropods. They also include some of the most spectacular living seashells such as cowries, murex, volutes and tritons. Geologically, gastropods are probably one of the least studied groups of molluscs. The reasons are twofold, the first being that they are not generally used to determine stratigraphic age, and secondly they are seldom well preserved as fossils, except in Tertiary deposits. The reason for their poor preservation is down to shell composition. The shells are generally aragonite, an unstable form of calcium carbonate that is easily dissolved or altered during burial.

44

Maclurea / Gastropod / Cambro-Ordovician

Phylum	Mollusca
Class	Gastropoda
Gen et sp	*Maclurea peachii*
Locality	Durness, Sutherland
Age	Cambro-Ordovician
Stratigraphy	Durness Limestone

This is not a pretty fossil – no more than the spiral cross-section of a gastropod. Technically it does not merit being given a generic, let alone a specific name. This specimen is just not well enough preserved to show details needed for naming a fossil. However, it is

44 The spiral is a cross-section of the gastropod *Maclurea* in Durness Limestone.

included as typical of fossils you might see in the Durness Limestone over its outcrop area from Skye to Durness.

This particular specimen is also interesting in that it was collected by Charles Peach, who was the first person to report fossils from the Durness Limestone in 1854. This find prompted Murchison to revisit the NW Highlands and revise some of his views on the age of the rocks in the area. The name *Maclurea peachi* was given to the fossil by Salter, who recognised the connection between the fauna of the Durness Limestone of Scotland and similar limestones in North America. This connection was greatly strengthened by the finding, many years later, of the trilobite *Olenellus* (*see* 58 below) in the Fucoid Beds beneath the Durness Limestone. Hence this rather poor specimen is historically interesting as part of the evidence that eventually led to the solution of the age and structure of the rocks of the NW Highlands by Peach (Ben) and Horne in the Geological Survey Memoir of 1907.

The specimen illustrated was clearly labelled after Salter described the fossil; this seems to imply that the printed label is later, but it is in Charles Peach's handwriting, as is the locality label.

45

Euphemites / Gastropod / Carboniferous

Phylum	**Mollusca**
Class	**Gastropoda**
Gen et sp	***Euphemites urii*** **(Fleming)**
Locality	**Roscobie, Fife**
Age	**Early Carboniferous**
Stratigraphy	**Lower Limestone Formation**

This small gastropod is a member of a primitive group called Bellerophontids, which had planispiral shells with spiral ribs and show bilateral symmetry; they are known from Cambrian to Triassic time. Most gastropods have helical shells and do not have bilateral symmetry. *Euphemites* is found along with brachiopods (e.g. *Spirifer, Productus*), bivalves, and crinoids in marine shales and limestones of the Carboniferous of Fife. They were part of a fauna that flourished in shallow seas that covered the area following rises in sea level that drowned the top of a large delta.

The Bellerophontids are thought to be primitive, and may be related to a group of molluscs called Monoplacophora, which are also bilaterally symmetrical and show some evidence of segmentation in the internal organs and muscles, maybe reflecting an origin of molluscs from a segmented ancestor such as a worm.

45 Views of the bellerophontid gastropod *Euphemites*.

ECHINODERMS

The Echinodermata, meaning 'spiny skinned', includes many familiar animals of the seashore including sea urchins, starfish, brittlestars, sea cucumbers, and sea lilies. Despite the names they have nothing to do with fish, vegetables or flowers. The most common found as fossils are the sea urchins; they have been most abundant from the Mesozoic to the present. However, they are not common in Scotland, and the classic sea urchins found in the chalk of England are only found in flints transported onshore in the ice age from rocks that occur on the seabed offshore.

On the other hand, Scotland has some excellent fossil starfish in thin 'starfish beds' that have been found in the Ordovician of the Girvan area and Silurian of the Pentland Hills.

Crinoids, or sea lilies, have long stems made of individual segments like small barrels that are called ossicles, on top of which is a cup, the calyx, with arms arranged in fivefold symmetry. The arms were used for gathering food, and passing it to the mouth at the top of the calyx. Stem fragments of crinoids are so abundant that they are the main component of some Carboniferous limestones, giving them the name 'crinoidal limestone'.

There are also several very obscure groups of animals within the phylum Echinodermata that are difficult to interpret. Several of these have been found in the Ordovician 'Starfish Bed' in the Girvan area. The primitive sea urchin *Aulechinus* is one example, and *Dendrocystoides* is included as an example of a group now thought to be a possible link between the echinodermata and chordates.

46

Aulechinus / Sea Urchin / Ordovician

Phylum	Echinodermata
Class	Echinoidea
Gen et sp	*Aulechinus grayii*
Locality	South Threave, Girvan, Ayrshire
Age	Ordovician, Ashgill
Stratigraphy	Starfish Bed, South Threave Fm., Drummock Group

Aulechinus is one of the oldest animals that can be recognised as a sea urchin. The test or 'shell' was made of numerous calcite plates that formed a flexible globular shape. This means that the plates were not rigidly fixed together as in modern sea urchins, so after death and decay the whole structure would have fallen apart. Thus the rare examples found in the 'Starfish Bed' were probably buried alive in a sediment flow. The test has a typical 5-fold symmetry with five narrow bands called ambulacra in which the plates had pores, indicating that it had tube feet. The ambulacra meet in a small disc at the apex with plates typical of echinoids, apart from only having one, rather than five, genital plates. It does not appear to have had spines, which are a feature of modern sea urchins. The specimen shows *Aulechinus* with two ambulacra preserved, in association with brachiopods.

Sea urchins are rare in the Palaeozoic, with fewer than 40 genera described worldwide. However, they became very common in Jurassic times and continue to the present day as an abundant element of marine faunas.

46 The primitive Ordovician echinoid *Aulechinus* (centre) in association with bivalves and brachiopods.

47

Holaster / Sea urchin / Cretaceous

Phylum	**Echinodermata**
Class	**Echinoidea**
Gen et sp	***Holaster subglobosus***
Locality	**Moss of Cruden, near Peterhead, Aberdeenshire**
Age	**Late Cretaceous**
Stratigraphy	**Buchan Ridge Formation (flint derived from Cretaceous chalk)**

This fossil is a flint cast of a sea urchin from the Cretaceous chalk. Thus it is surprising to find it in northern Scotland, rather than in the south of England, where the chalk is exposed and flint casts of urchins are common. Beneath the peat on the Moss of Cruden is a layer of pebbles that includes flints that were derived from the chalk. The origin of these flints and the possibility that the Late Cretaceous chalk sea covered part of NE Scotland is discussed in the entry for the sponge *Ventriculites*.

Sea urchins were common in the chalk seas, and most had small spines and burrowed in the fine carbonate mud in search of food particles. It seems that the chemical conditions within the shell after the death of the animal were suitable for the deposition of silica, now preserved in the form of flint. The shell is of calcite, which later dissolved to leave a cast of the inside surface of the urchin shell. Thus the external features of the shell, such as the attachment points for the spines, are not preserved.

The presence of this urchin along with other fossils such as bivalves clearly demonstrates the late Cretaceous age of the flint.

47 Flint cast of the Cretaceous echinoid *Holaster.*

48

Dendrocystoides / 'Primitive echinoderm' / Ordovician

Phylum	**Echinodermata, subphylum Homalozoa**
Class	**Homoiostelea**
Gen et sp	***Dendrocystoides scoticus***
Locality	**South Threave, Girvan, Ayrshire**
Age	**Ordovician, Ashgill**
Stratigraphy	**Starfish Bed, South Threave Fm., Drummock Group**

This very strange animal has been the subject of much academic study, with different proposals put forward on the zoological relationships of the animal, and interpretation of the anatomy. There are even differences of opinion as basic as which orifice is the mouth and which

48 *Dendrocystoides*, a primitive member of the Echinodermata preserved with the starfish *Dreparaster.*

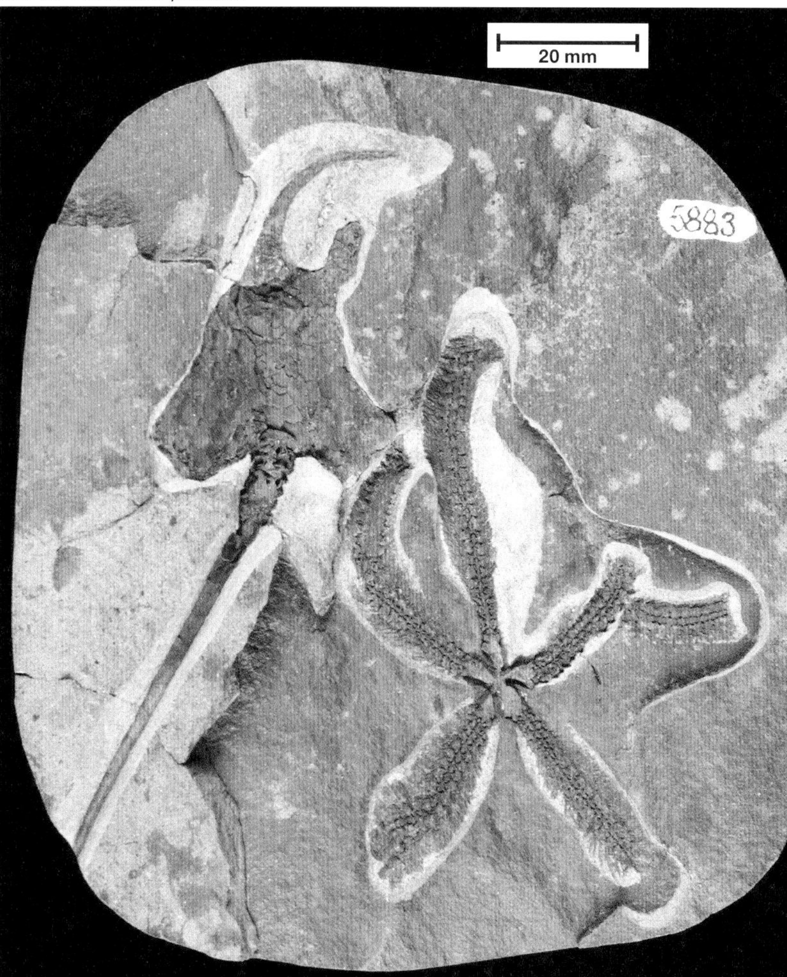

the anus. Traditionally it is classified in the Echinodermata, and will be found there in most text books (*see* Clarkson 1998). It has been suggested by Jefferies (1986) that it has strong similarities with *Cephalodiscus*, which is a Hemichordate. This means that it may be a link between Echinodermata and Chordates, and Jefferies places it in a Class Soluta. However, this idea has not become generally accepted amongst those working on early chordates (*see* Benton 2009).

Dendrocystoides lacks the fivefold radial symmetry of most Echinodermata and may have lain flat on the sea floor; however, jointed 'arms' and 'stems' of echinodermata are remarkably flexible and the 'tail' of *Dendrocystoides* may have dug into the sediment, allowing the animal to raise itself off the sea floor.

In this specimen an excellent complete *Dendrocystoides* lies beside a similarly well-preserved starfish (*Dreparaster grayae*). Both were probably buried alive, which would account for the fully articulated preservation of such delicate animals. The following starfish *Stenaster* was also preserved in the same way.

49

Stenaster / Starfish / Ordovician

Phylum	**Echinodermata**
Class	**Asteroidea**
Gen et sp	***Stenaster obtusus***
Locality	**South Threave, Girvan, Ayrshire**
Age	**Ordovician, Caradoc**
Stratigraphy	**Starfish Bed, South Threave Fm., Drummock Group**

Stenaster is the starfish most commonly found at the South Threave locality. Most are about a centimetre across, and have been decalcified during preservation. The starfish was made up of calcite plates that gave it a fairly rigid structure, but the plates have been dissolved to leave holes in the rock. Thus the specimens are now hollow, but detail of the surface ornament is frequently well preserved as an impression on the rock matrix.

49 The small starfish *Stenaster*, preserved as a hollow mould.

50

Lepyriactis / Starfish / Ordovician

Phylum	Echinodermata
Class	Asteroidea
Gen et sp	***Lepyriactis nudus*** **Spencer**
Locality	Gutterford Burn, Pentland Hills
Age	Silurian, Llandovery
Stratigraphy	Reservoir Formation

This fine starfish is from a thin sandstone bed in the Silurian of the North Esk Inlier known as the 'Starfish Bed'. This seems to imply that starfish are common; they are not! Beds with such fine starfish are rare, and usually not laterally extensive. In this specimen the calcite plates of the starfish have been dissolved to leave holes that are stained by brown iron hydroxide. The depositional environment is probably part of a submarine fan, and the starfish may have been buried by the deposits of turbidity currents sweeping over the fan surface. The starfish bed was excavated in the late nineteenth century. This specimen was collected before 1897, the date it was acquired by RMS. From the time of the first discovery of fossils in the Silurian of the North Esk Inlier in 1836, there have been many studies on different aspects of the fauna that continue to the present day. In such areas of poor exposure there are many new finds to be made, and many localities would repay further excavation. Clarkson (1985) gives a brief account of the history of research in the area.

50 The starfish *Lepyriactis.*

51

Cupulocrinus / Crinoid / Ordovician

Phylum	Echinodermata
Class	Crinoidea
Gen et sp	***Cupulocrinus heterobrachialis***
Locality	South Threave, Girvan, Ayrshire
Age	Ordovician, Caradoc
Stratigraphy	Starfish Bed, South Threave Fm., Drummock Group

This is a small crinoid showing typical Starfish Bed preservation with the calcite plates dissolved out, and the crinoid preserved as a hollow mould. Part of the stem can be seen, together with a small elongate calyx and arms. The Ordovician was a time of great radiation in crinoids, and the Starfish Bed has produced many forms known from nowhere else. The richness of this locality is due to the sudden burial of the living fauna, also including starfish and forms such as *Dendrocystoides,* all of which can be found in close association within the Starfish Bed. About 50 British Ordovician crinoids are known, of which about a third are only known from this locality. Many were collected by Mrs Gray, and Ramsbottom (1961) incuded some in his monograph of British Ordovician crinoids, including *Cupulocrinus.*

51 *Cupulocrinus*, a small crinoid preserved in sandstone.

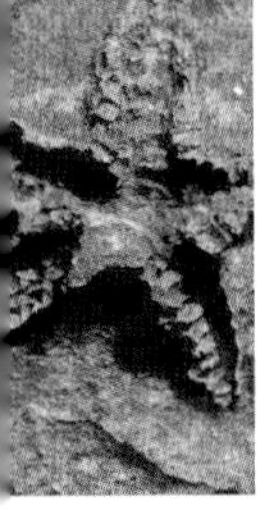

52

Crinoid ossicles / Crinoid / Carboniferous

Phylum	**Echinodermata**
Class	**Crinoidea**
Gen et sp	**Crinoid ossicles from more than one genus**
Locality	**Clatteringwell Quarry, Bishop Hill, Kinross**
Age	**Lower Carboniferous**
Stratigraphy	**Charlestown Main Limestone, Lower Limestone Formation.**

Crinoid ossicles, segments of the stems of crinoids, are some of the most common Scottish fossils, and are abundant in the marine limestones of the Lower Carboniferous. The examples shown were picked up on the tips of the long disused Clatteringwell Quarry on the top of Bishop Hill, part of the Lomond Hills, overlooking Loch Leven near Kinross. Stems of these Carboniferous crinoids are circular or oval in cross-section. Later in Jurassic times some crinoids had stems that are pentagonal, or a five-pointed star in cross-section. Crinoid stems show many methods of articulation and were flexible in life, the animal having control over flexing of the stem. Many stems also had side branches called cirri; the attachment points to the main stem occur on specialised segments of stem.

In early Carboniferous times sea level fluctuated as ice caps at the poles expanded and contracted due to long-term climatic fluctuation. During times of high sea level the Midland Valley region of Scotland was flooded by a clear tropical sea, and marine life flourished with diverse faunas of corals, brachiopods and crinoids. The crinoids were so abundant that in some areas they formed underwater forests with each crinoid cup at the top of a very long stem. It seems likely that the forms that could raise themselves highest had the best feeding, rather like trees in the forest growing tall to compete for light.

On death the stems broke up easily and were transported and sorted by currents to produce thick beds made mainly of crinoid stem debris. After burial this debris was cemented with calcite and became crinoidal limestone. It is very difficult to find any evidence of the cups of the crinoids at Clatteringwell Quarry, and it is possible that the cups either disintegrated into isolated plates, or that they could detach themselves from the stem and drift to a more favourable location, and grow a new stem. Some modern crinoids have this ability. The animals clearly invested a lot of effort into depositing the calcite required for these stems, which are enormous when compared with the body of the animal situated in the cup on top of the stem.

52 Crinoid stems of several types weathered out of muddy limestones and shales.

53

Woodocrinus / Crinoid / Carboniferous

Phylum	**Echinodermata**
Class	**Crinoidea**
Gen et sp	***Woodocrinus gravis***
Locality	**Inverteil, Fife**
Age	**Carboniferous**
Stratigraphy	**Lower Limestone Formation**

Crinoids, often called 'sea lilies', are animals in which the body and arms used to gather food are elevated from the sea floor on a stem. This enables the animal to filter-feed more effectively. It is rather like an inverted starfish on a stalk, but not all crinoids have stalks; some are free swimming.

This very fine crinoid with part of the stem and the crown is from a marine band rich in crinoids within the Lower Limestone Group. James Wright (1878–1957) collected and studied the crinoids of the Carboniferous. He wrote his first paper in 1911 and his major works were the monograph *British Carboniferous Crinoids* (1952–60), and *Scottish Carboniferous Crinoids* (1939). His collection of crinoids passed to the Royal Museum of Scotland in 1958, and contains the Type material of numerous Scottish crinoids and remains a valuable reseach resource. Complete crinoids such as this are rare, since following death of the animal, the skeleton of calcite plates breaks up very easily. This individual from a shaly bed was probably choked and buried alive by mud entering the shallow Carboniferous sea from a delta at a time of flood.

Wright carefully extracted the fossil from the matrix and mounted it on card. The card reveals that this specimen is a 'Paratype', part of the Type material on which the species was founded by Wright. Thus the card gives references to publications in which it was described. Wright was a palaeontologist who documented his specimens carefully. The card shows the number of the specimen in the 'Jas Wright Collection', and also the 1958 acquisition number to the RMS collection.

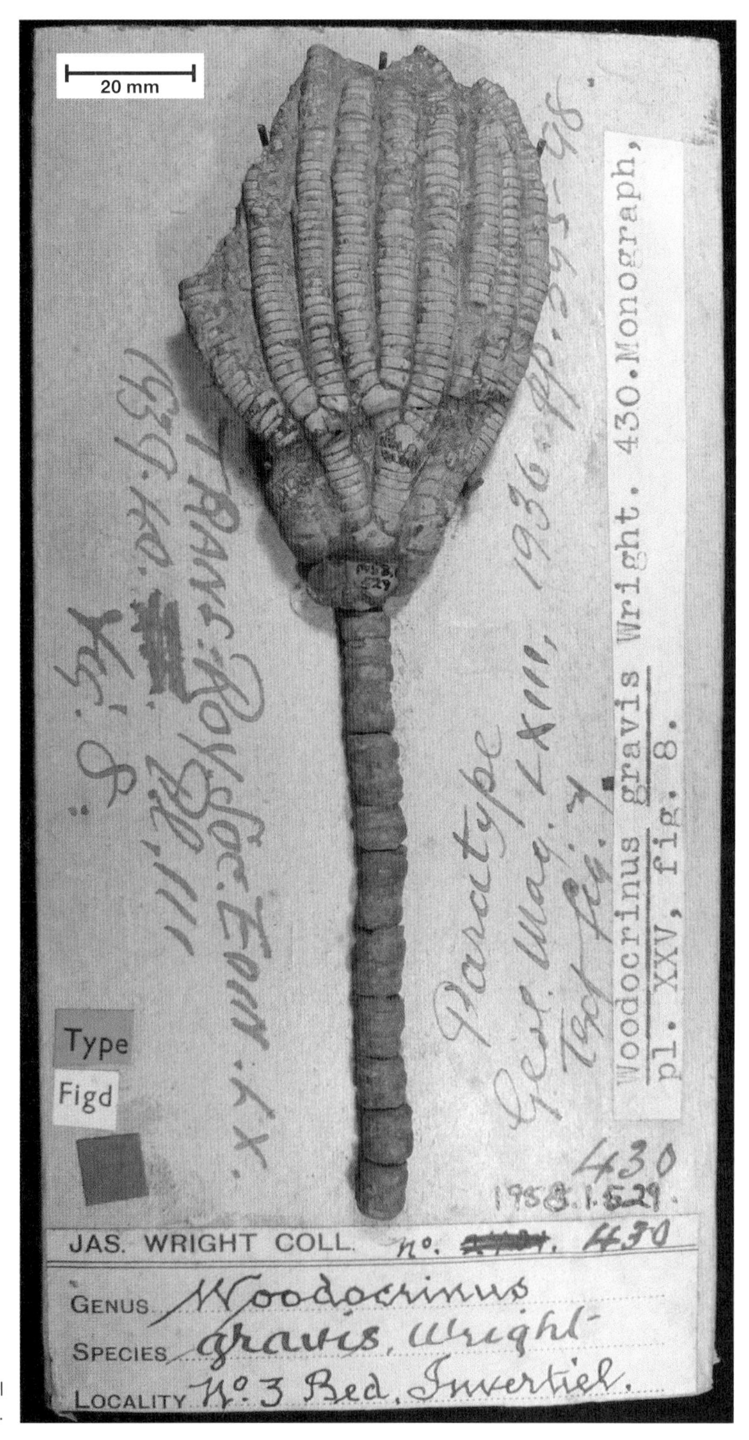

53 Type material of *Woodocrinus* in the National Museums Scotland collection.

54

Tubulusocrinus / Crinoid / Carboniferous

Phylum	**Echinodermata**
Class	**Crinoidea**
Gen et sp	***Tubulusocrinus doliolus***
Locality	**St Andrews, Fife**
Age	**Carboniferous**
Stratigraphy	**Pittenweem Formation, Strathclyde Group**

This crinoid has had its name changed three times. Originally described by James Wright (1934) as *Hydriocrinus*, it became *Ulocrinus* in 1936, *Ureocrinus* in 1948 and was given its present name by Kammer and Ausich (2007). Name changes are inevitable as perceptions change on classification and on the significant differences that merit a new generic name. What has not, and cannot, be changed is the specific name, unless it is discovered that it had been described under another name prior to 1934.

This fine crinoid shows a small section of the stem, a very well-preserved calyx on which the individual plates that form the calyx can be clearly seen, and three (of five) long delicate arms. In their study of the illustrated specimen, which is the Type specimen, Kammer and Ausich recognised that soft tissue of an anal tube has been preserved. On the specimen this is the short reddish tube protruding from the calyx between the arms at the left-hand side in the picture. This tube was used to ensure that waste products did not mix with food gathered by the arms, acting rather like an extractor fan. It was this feature (that Wright had not recognised), that led to a change of name. The crinoid lies on top of a thin limestone which is composed mainly of crinoid debris. This thin bed is known as the 'Encrinite Bed', and occurs within the Carboniferous succession on the shore near St Andrews a few hundred metres from the famous 'Rock and Spindle' volcanic vent. The Encrinite Bed is part of a marine band (Marine Band II of Macgregor, 1973, and the Witch Lake Marine Band of Forsyth and Chisholm, 1977). The marine band consists of about 3.6m of grey shale with brachiopods, gastropods and bivalves. The Encrinite Bed is a limestone up to 15 cm thick within this shale. It appears that crinoids flourished briefly, and a bed of crinoid debris accumulated at a time of low mud supply to the sea floor. The well-preserved crinoids were probably killed and instantly buried by an influx of mud, and are thus found lying on top of the bed. The fact that the delicate anal tube is preserved supports this interpretation. The general environment was deltaic, but at times of high sea level the delta top was invaded by the sea and became a shallow marine area. As sea level fell, the delta built out into the sea and covered the shallow marine area with delta muds and sands. Thus changing sea levels, combined with general subsidence, produced a cyclic succession of marine and deltaic strata.

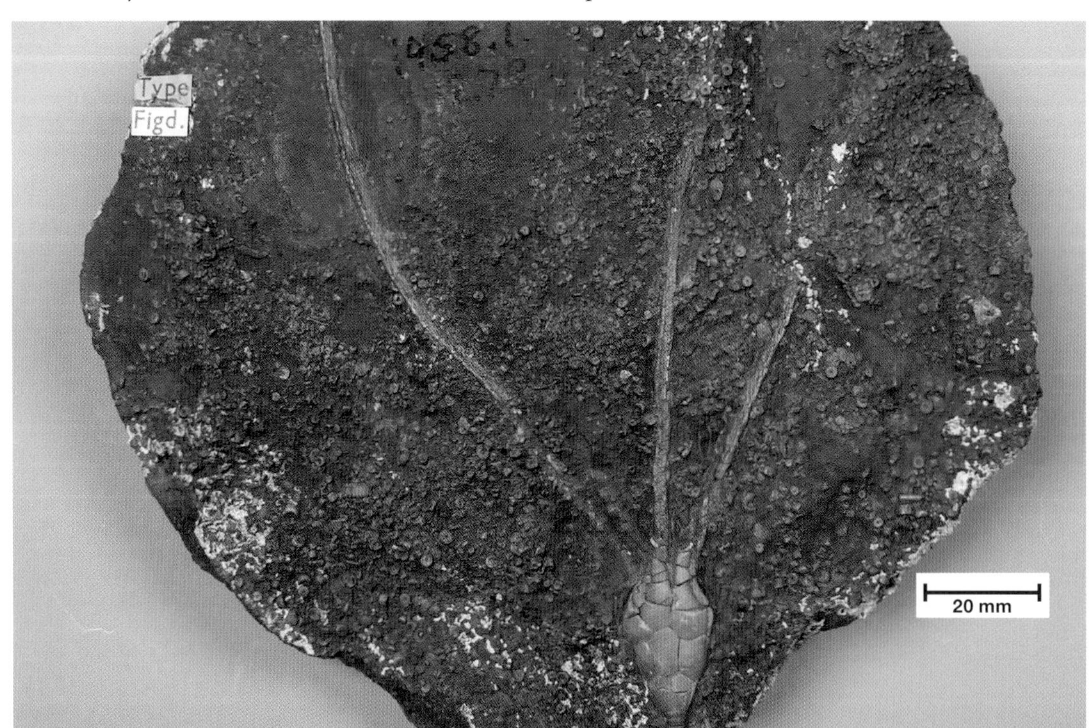

54 *Tubulusocrinus*, preserved on the top of a thin limestone bed, and probably killed by rapid deposition of mud.

GRAPTOLITES

Graptolites were colonial animals that secreted an organic structure with branches (stipes) that bore cups (thecae) that held the individuals that made up the colony (rhabdosome). Each colony developed from a single individual in a distinctive manner, and the numerous genera and species are distinguished on the basis of colony and theca shapes. The tubular parts of the colony contain collagen fibres, and are built of half-rings that interlock with a zig-zag suture. Graptolites were abundant in the oceans of the Ordovician and Silurian periods, and evolved very rapidly. They floated within the ocean water and when they died the organic colonies sank to the bottom and are commonly found preserved in black shales of deep water origin.

Because they evolved rapidly they make ideal 'zone fossils', useful for correlating rocks of the same age in different areas. Charles Lapworth was a pioneer in creating graptolite zonal schemes, basing much of his analysis on careful study of the successive graptolite faunas found in the Ordovician and Silurian rocks at Dob's Linn near Moffat (Lapworth 1878). The cottage he lived in whilst working in the area is known as 'Lapworth's Cottage'. Advances continue in graptolite biostratigraphy in Scotland, as shown by the series of papers in the Scottish Journal of Geology introduced by Stone *et al.* (2003). Whilst Lapworth made the great breakthrough, others continue to refine zonal schemes that unravel the biostratigraphy of the Southern Uplands and Girvan successions.

The relationships of graptolites were long disputed, frequently being grouped with corals despite their different chemistry. They are now regarded as hemichordates following comparisons with a modern colonial organism called *Rhabdopleura*, which constructs tubes in a similar way to graptolites, and is chemically similar.

Most graptolites are found crushed in black to grey shale or slate, and are difficult to interpret, but occasionally they are preserved in 3D in limestones from which they can be extracted using acids. These 3D specimens have provided most of the information on the detailed structure of graptolites.

55

Dicellograptus / Graptolite / Ordovician

Phylum	Hemichordata
Class	Graptolithina
Gen et sp	*Dicellograptus*
Locality	Dob's Linn, Moffat
Age	Late Ordovician
Stratigraphy	Hartfell Shale Formation, Moffat Shale Group

There are several species of *Dicellograptus* that are distinguished by details of the colony form and the thecae. The general pattern is of two stipes that diverge upwards from the sicula – the structure formed by the first individual of the colony. In some forms the stipes are straight, in others they are twisted. The different forms existed for short periods of time, making them useful for correlation.

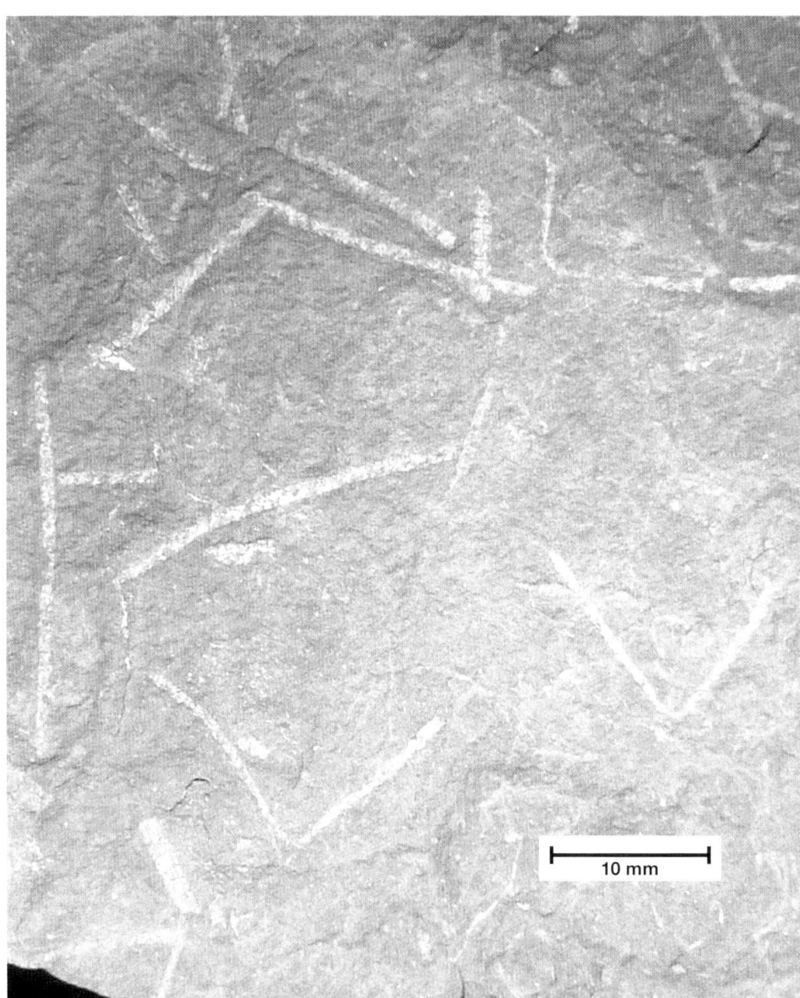

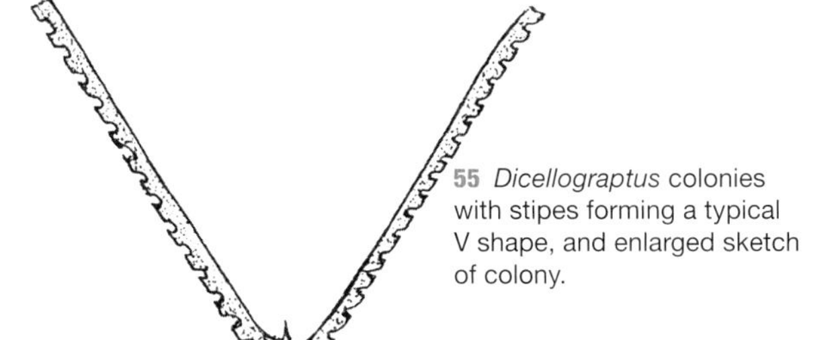

55 *Dicellograptus* colonies with stipes forming a typical V shape, and enlarged sketch of colony.

56

Climacograptus / Graptolite / Silurian

Phylum	Hemichordata
Class	Graptolithina
Gen et sp	*Climacograptus bicornis*
Locality	Dob's Linn, Moffat
Age	Early Silurian
Stratigraphy	Birkhill Shale Formation, Moffat Shale Group

Climacograptus is a biserial graptolite, with thecae alternating on both sides of the stipe. The colony grows upwards from the sicula, and a long thread may extend from the top of the colony. The illustrated species has distinctive 'horns' that are reflected in the specific name. Graptolites were planktonic, and many of the structures seen in colonies were probably buoyancy aids, controlling the depth at which the colony floated in the water. The thecae of *Climacograptus* have characteristically straight outer margins, and narrow apertures.

56 Biserial *Climacograptus* colonies with double 'horns', hence the specific name *bicornis*, together with an enlarged sketch of colony.

57

Monograptus / Graptolite / Silurian

Phylum	**Hemichordata**
Class	**Graptolithina**
Gen et sp	***Monograptus priodon***
Locality	**Grieston Quarry, Innerleithen, Peeblesshire**
Age	**Early Silurian**
Stratigraphy	**Gala Group**

Monograptus priodon has a long, straight, robust stipe with thecae along one side. The thecae are hooked in shape so that the apertures point back towards the origin of the stipe. This was a large graptolite, with some stipes over 25cm long. There are several graptolite-rich shale beds in Greiston Quarry with well-preserved specimens, but in some beds the graptolites have been broken during transport by sediment-laden currents. Evidence of graptolite transport is provided by sediment-filled impact impressions of graptolites on the bases of some beds.

This was possibly the second graptolite locality to be recorded in Scotland, being mentioned by James Nicol in his *Geology of Scotland* (1844), and in an essay on 'The Geology of Peeblesshire' (1843). Nicol lived at Innerleithen at that time, but was subsequently Professor of Natural History at Aberdeen University, and played an important role in interpreting the structure of the NW Highlands.

A specimen of this graptolite was found in the Hugh Miller Collection with a label indicating that it came from the Dalradian Macduff Slates in Banffshire. The specimen was given to Hugh Miller by a Dr Emslie of Banff. The said doctor had studied geology under James Nicol at Aberdeen, and modern mineral and chemical investigation soon revealed that the rock matched that of Greiston Quarry, and not the Dalradian of Banffshire. A Silurian graptolite from the Dalradian is clearly impossible, given modern knowledge that the Dalradian is intruded by igneous rocks of Ordovician age, but in Hugh Miller's day the slates of Banffshire were not seen as being significantly different from those of the Southern Uplands. So the good Dr Emslie told a small fib for the time to Hugh Miller, but he has been found out by modern science.

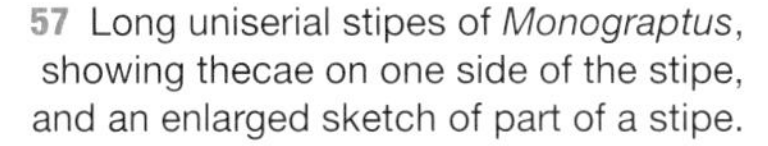

57 Long uniserial stipes of *Monograptus*, showing thecae on one side of the stipe, and an enlarged sketch of part of a stipe.

ARTHROPODS

Arthropods have external skeletons and jointed legs, and are the most numerous and diverse group of animals on Earth. They include the insects, crabs, barnacles, shrimps, millipedes, centipedes, spiders, scorpions and horseshoe crabs. Many of these groups have a long history and some of the earliest representatives are found in Scotland.

The best-known extinct group of arthropods is the trilobites. As the name suggests, the trilobite is divided into three parts; a head (cephalon), segmented body (thorax) and tail (pygidium). They are also divided in three laterally with a central axis, flanked with flatter lobes. They occur in rocks of marine origin from the Cambrian to the Carboniferous in Scotland. Particularly fine specimens occur in the Ordovician of the Girvan area, and the Silurian of the Pentland Hills.

Near Lesmahagow and in the Pentland Hills Silurian strata have yielded a rich fauna of arthropods called eurypterids. One, *Slimonia*, was over a metre long. Some forms such as *Erretopterus* have pincers and appear to have been predators. They are frequently associated with the pod-shrimp *Ceratiocaris.*

The Early Devonian Rhynie chert in Aberdeenshire contains a wide variety of different freshwater and terrestrial arthropods, with insects, mites, crustaceans, millipedes, centipedes, and representatives of several extinct groups. Several of these animals are the oldest known fossils of the group.

In the Carboniferous of the Midland Valley there are some remarkable beds with myriads of very well-preserved shrimps such as *Tealliocaris* that lived in a lagoon close to the sea – even details of the delicate legs and antennae are seen.

Fossil lobsters that lived in burrows on the sea floor are found in the Jurassic rocks of Skye. They are not always well preserved, and many specimens probably moulted shells left in burrows. Also on Skye, organic-rich lagoonal shales in the Great Estuarine Group contain masses of ostracods and the branchiopod crustacean *Estheria*.

58a *Olenellus* headshield from the Lower Cambrian, NW Highlands.

58

Olenellus / Trilobite / Cambrian

Phylum	**Arthropoda**
Class	**Trilobita**
Gen et sp	***Olenellus lapworthi***
Locality	**Kinlochewe, Sutherland**
Age	**Early Cambrian**
Stratigraphy	**Fucoid Beds, An-t-Sron Formation**

Olenellus is a primitive type of trilobite, having many segments in the body, a small tail, and a segmented axis in the head flanked by long crescentic eyes. It is found in the Fucoid Beds of the Lower Cambrian succession in the NW Highlands; the word 'fucoid' means seaweed, and refers to marks in the rock once thought to be seaweed impressions, but now recognised as burrows. The animal lived on the sea-floor, ploughing through the surface sand and mud. Bilobed trackways with scratches made by the claws on the legs are also found in the Fucoid Beds. This is the fossil that clearly established the age of these rocks, by comparison with similar trilobites in the Appalachians of North America.

Olenellus was figured in the famous Geological Survey Memoir on *The Geological Structure of the North-West Highlands of Scotland* (Peach *et al.* 1907), but was first discovered by Lapworth (1888) and described by Peach (1894), and more recently by Cowie and McNamara (1978). The similarity of the rocks and fossils of the NW Highlands with those of the Appalachians of Eastern Canada and the USA was one piece of early evidence that the areas had once been closer together. We now know that they have been separated by the opening of the Atlantic Ocean, a process that started only 65 million years ago.

Complete specimens of this fossil are exceptionally rare, but the distinctive headshield is more common. As arthropods, trilobites had to moult periodically in order to grow, and most fossils of trilobites that are found are probably parts of the moulted shell.

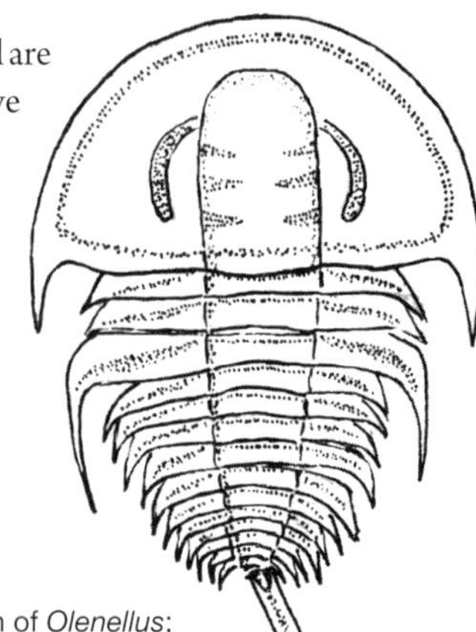

58b Reconstruction of *Olenellus*; complete specimens are rare.

59

Lonchodomas / Trilobite / Ordovician

Phylum	**Arthropoda**
Class	**Trilobita**
Gen et sp	***Lonchodomas macullumi***
Locality	**Dow Hill, Girvan, Ayrshire**
Age	**Ordovician**
Stratigraphy	**Lower Balclatchie Group**

Lonchodomas was a strange beast! It had no eyes and the head was adorned by three long spines. The genal spines extended far beyond the tail, and there was a sharp spine on the front of the head. The flexible segments of the body allowed the animal to roll up and protect the underside. The specimen illustrated is rolled up with the tail tucked under the head; one of the long genal spines is preserved. Blind trilobites are frequently interpreted as burrowing forms, or dwellers in deep water. *Lonchodomas* can hardly have been a burrower, and was most probably pelagic, drifting in the oceans. The spines may have served a dual function for flotation and defence – it would not have been easy to swallow.

59 *Lonchodomas*, a blind trilobite with long genal spines.

60

Proetidella / Trilobite / Ordovician

Phylum	**Arthropoda**
Class	**Trilobita**
Gen et sp	***Proetidella girvanensis***
Locality	**South Threave, Girvan**
Age	**Ordovician**
Stratigraphy	**Starfish Bed, South Threave Fm., Drummock Group**

The 'Starfish Bed' at Threave Glen has been a magnet for fossil collectors for decades, and is the source of many beautiful specimens, particularly those in the Hunterian Museum in Glasgow. These specimens are from the famous Gray Collection, formed by Mrs Robert Gray and acquired by the Hunterian Museum in 1866. Other collections of the Gray family went to Edinburgh and the Natural History Museum in London. The fauna preserved in the Starfish Bed includes starfish, primitive echinoderms, trilobites, brachiopods, gastropods and many others. The rock is a greenish sandstone, and most of the fossils have been decalcified, meaning that the actual shell has been dissolved, to leave a fine impression of the animal. The fragment illustrated contains a complete trilobite, *Proetidella*, which is possibly the most common trilobite present. The same small specimen also has two gastropods, a brachiopod and part of a starfish. This is probably a 'displaced fauna', where the animals were caught in a sediment flow and transported into deeper water. Thus the fauna was buried alive, accounting for the presence of complete trilobites and starfish. Under normal circumstances they would have broken up after death, or been scavenged by other animals.

60 *Proetidella* with a gastropod, brachiopod and part of a starfish in a small piece of the Starfish Bed.

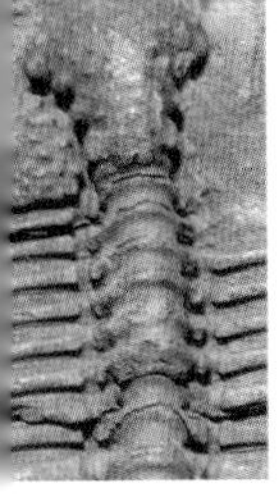

61

Calyptaulax / Trilobite / Ordovician

Phylum	**Arthropoda**
Class	**Trilobita**
Gen et sp	***Calyptaulax brongniartii***
Locality	**Dow Hill, Girvan, Ayrshire**
Age	**Ordovician, Caradocian**
Stratigraphy	**Lower Balclatchie Group**

This trilobite was originally placed in the genus *Phacops,* and shows the typical large schizochroal eyes with about 150–200 individual circular lenses in each eye. This type of eye is unique to phacopid trilobites and contrasts with the normal trilobite compound eye in which all the lenses are in contact and tend to have hexagonal shapes. By plotting the orientation of each lens in schizochroal trilobite eyes, Clarkson (1966) showed that they had very good all-round vision in the horizontal plane, but that the field of vision did not extend more than about 20 degrees above the horizon. Hence this trilobite could hunt and see potential predators as it sat on the sea floor, but was vulnerable from above. It is possible that the rapid evolution and increase in abundance of fish in the Palaeozoic was a significant factor in the extinction of the trilobites. *Calyptaulax* occurs at several localities in the Girvan area, also near Abington, and at Pomeroy in Northern Ireland. This species shows some variation between different localities in the number and arrangement of lenses in the eyes, but these are not considered important enough to merit different species names (Clarkson and Tripp 1982). The variations may be due to the evolution of the animal through time, or could also be a response to variable environmental conditions.

61a *Calyptaulax*, dorsal view of cephalon.

61b *Calyptaulax*, side view with cephalon bent at 90 degrees to body.

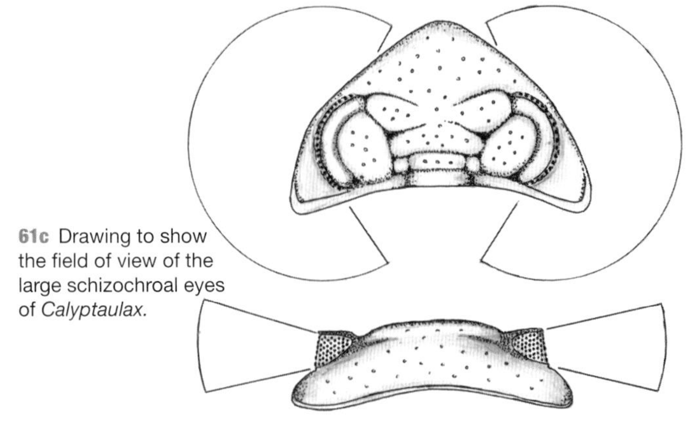

61c Drawing to show the field of view of the large schizochroal eyes of *Calyptaulax.*

62

Encrinurus / Trilobite / Silurian

Phylum	**Arthropoda**
Class	**Trilobita**
Gen et sp	***Encrinurus pagei***
Locality	**Henshaw Burn, near Carlops, Edinburgh**
Age	**Silurian, Upper Llandovery**
Stratigraphy	**Wether Law Linn Formation**

Encrinurus is commonly known as the 'strawberry-headed trilobite' on account of the pustular bumps that cover the head. The specimen is virtually complete and may have been caught in a sediment flow and transported from a shallow area to deep water. The transported fauna would have included empty shells of animals already dead, together with the living fauna. Storm action can stir up the sediment and start a sediment flow. The genus ranges from the Ordovician to Silurian periods, and some of the best-known specimens come from the Middle Silurian Wenlock Limestone of Dudley in Worcestershire. This specimen is slightly older, and is from rocks in the North Esk Inlier of Silurian rocks to the south of Edinburgh. This species was described by Haswell in 1865; he was one of a band of enthusiastic collectors in the Edinburgh Geological Society (Clarkson 1985).

The succession of this inlier shows a transition from rocks with a clearly marine fauna such as this trilobite, up into rocks with increasingly non-marine faunas, including primitive fish. It is thought that the basin in which these rocks were deposited was cut off from marine connections during the closure of the Iapetus Ocean, and marine faunas died out, to be replaced with forms that could tolerate the changing conditions.

Those who visit the Pentland Hills will find the illustrated guide to the Silurian fossils by Clarkson *et al.* (2008) essential for the identification of specimens.

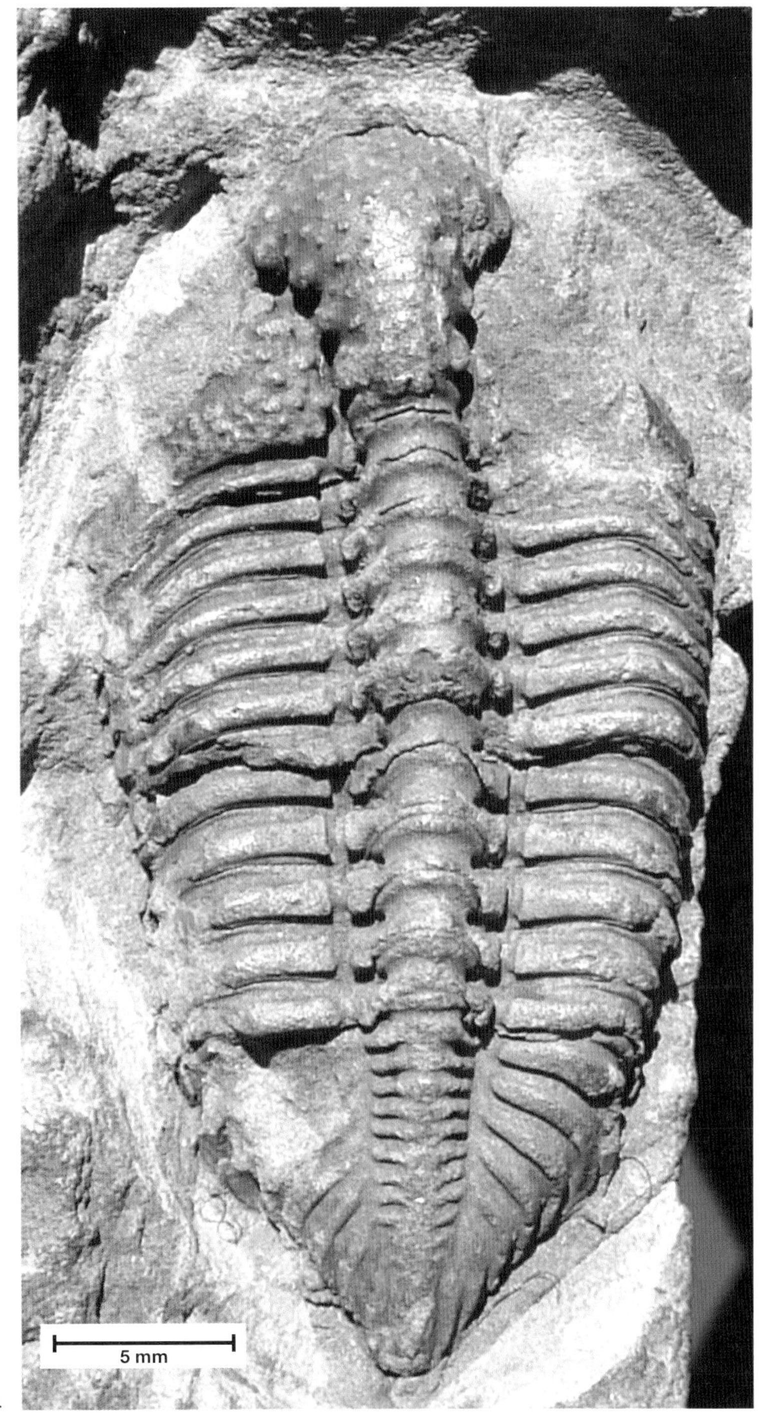

62 *Encrinurus*, a fine specimen from the Pentland Hills.

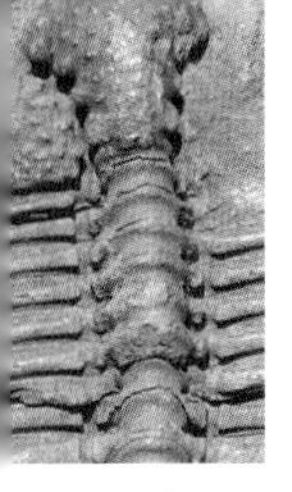

63

Erretopterus / Eurypterid / Silurian

Phylum	**Arthropoda**
Class	**Chelicerata, Merostomata**
Gen et sp	***Erretopterus bilobus***
Locality	**Logan Water, Lesmahagow**
Age	**Silurian, Llandovery**
Stratigraphy	**Blaeberry Formation**

The Silurian rocks of the Lesmahagow area are famed for the array of well-preserved eurypterid fossils brought to notice by Robert Slimon of Lesmahagow in 1855. Specimens can be seen in many museum collections, and *Erretopterus* is a typical representative. Complete specimens are rare; the example illustrated shows the head, segmented abdomen and tail. Remarkably, some of the appendages are preserved, including the broad paddle-like appendages used for swimming and one of the pincers, which protrudes at the front from below the head. *Erretopterus* was probably a carnivore preying on other arthropods, and possibly on some primitive fish.

All arthropods have to moult to grow larger, and most of the specimens found are probably parts of moults that were cast off by the animal. The eurypterids lived in a body of water that was initially marine, but was progressively cut off from the ocean in the Silurian as the Iapetus Ocean closed. Animals requiring marine conditions, such as brachiopods, died out, but a variety of arthropods such as *Errretopterus* flourished. As the water became fresh the arthropods and some primitive fish became the main components of the fauna.

63 The eurypterid *Erretopterus* from the Lesmahagow area, complete with claw-bearing appendage, and flattened swimming appendages at side.

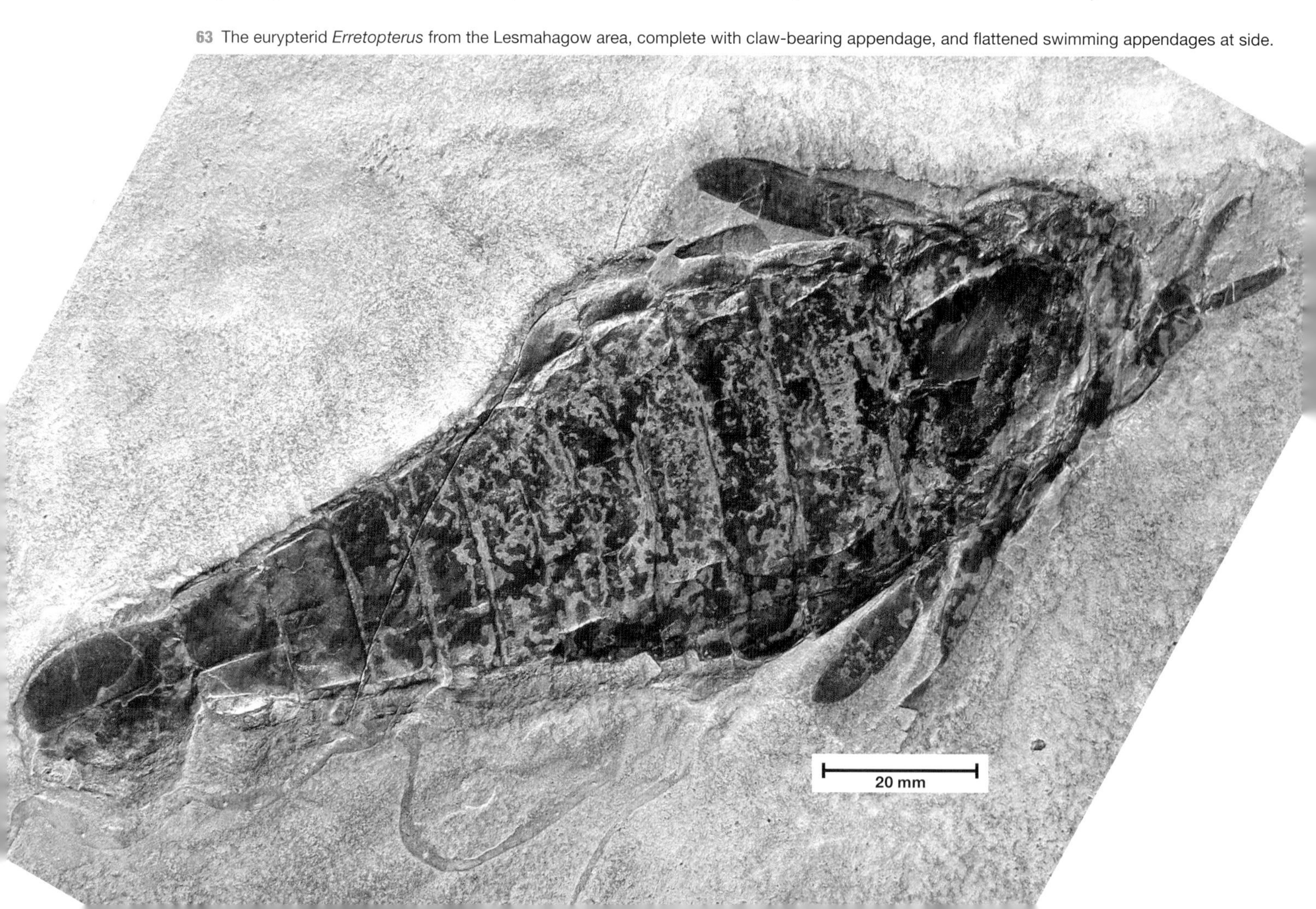

64

Pterygotus / Eurypterid / Devonian

Phylum	**Arthropoda**
Class	**Chelicerata, Merostomata**
Gen et sp	***Pterygotus anglicus***
Locality	**Carmyllie, Forfar, Angus**
Age	**Early Devonian**
Stratigraphy	**Arbuthnott Group, Lower Old Red Sandstone**

Pterygotus was a giant eurypterid that inhabited rivers and lakes in the Midland Valley in the Early Devonian. It grew to a metre and a half long, and was the largest animal, and possibly the top predator in the food chain. The most striking feature apart from size, is the pincers, or chelicerae, which resemble crab claws, but have sharper points and serrations that acted like scissors. It probably preyed on other arthropods and fish. With a flattened body it may have lurked at the bottom, waiting for prey to pass close enough to grab, or it might have cruised the surface in search of victims. Fragments of this animal can still be found in old quarry tips in the Forfar–Dundee region.

Most of the large complete specimens were found when quarrying was active in the Forfar area, and can only be seen in museums. Sadly, one fine specimen was sent to Germany by the Montrose Museum in exchange for a collection of European fossils – I am sure they assumed that more complete *Pterygotus* specimens would be found in the quarries. However, none came to the museum, and the quarries closed. After the Second World War, the museum asked if the exchange could be reversed, as they wished to concentrate on local geology. The reply was brief; the specimen had been destroyed by the RAF in a bombing mission.

64a *Pterygotus* reconstruction.

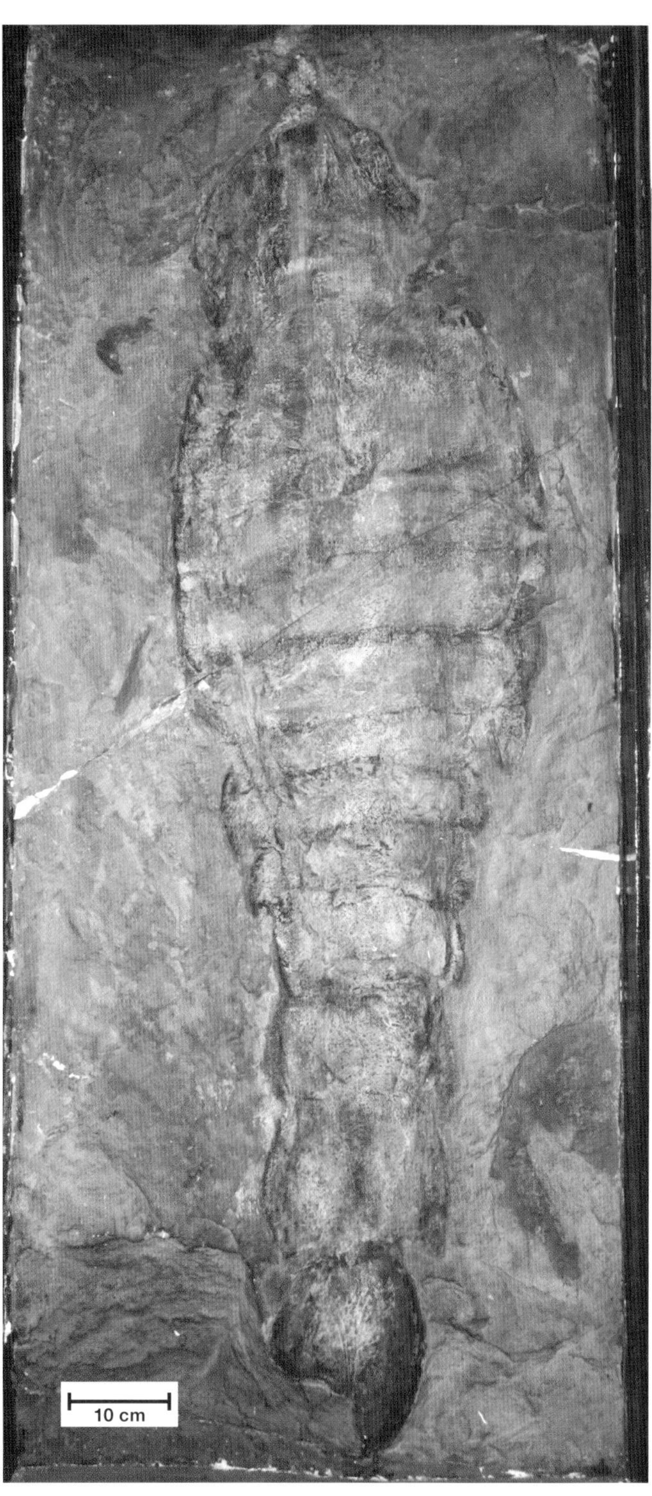

64b The giant eurypterid *Pterygotus* from the Lower Old Red Sandstone (plaster cast).

65

Pulmonoscorpius / Scorpion / Carboniferous

Phylum	**Arthropoda**
Class	**Scorpionida**
Gen et sp	***Pulmonoscorpius kirktonensis***
Locality	**East Kirkton Quarry, Bathgate, West Lothian**
Age	**Lower Carboniferous, Brigantian**
Stratigraphy	**East Kirkton Limestone**

Yes! This is a fossil and not a modern preparation. Scorpions are an ancient group dating from the Devonian, and this Carboniferous specimen from East Kirkton is a juvenile that has been extracted from finely laminated limestone by acid digestion of the rock. The specimen is from East Kirkton Quarry, famed for fossil tetrapods, but this specimen, described by Jeram (1994), is simply spectacular. The typical scorpion features of a narrow abdomen with a sting, four pairs of walking legs and pincers are clearly seen. The cuticle of the scorpion is acid-resistant, and with very gentle treatment the crushed scorpion floats free of matrix during acid treatment and can then be carefully mounted on glass. For his study of this arthropod Andrew Jeram obtained 16 complete or near-complete specimens, mostly of juveniles around 20mm long, but adults may have reached 700mm based on partial specimens. This is the best growth series known for a Palaeozoic scorpion.

The scorpions probably lived on the Carboniferous forest floor amongst plant litter, and the fossils represent individuals that were transported into a lake, probably by runoff following rain, and deposited in very low energy conditions. It seems that adults seldom met their end in this way, since most of the material is from juveniles. A 700mm long scorpion would have been able to tackle quite large prey items, and was probably a significant predator on the Carboniferous forest floor.

65 Remarkable preservation of the Carboniferous scorpion *Pulmonoscorpius* from East Kirkton.

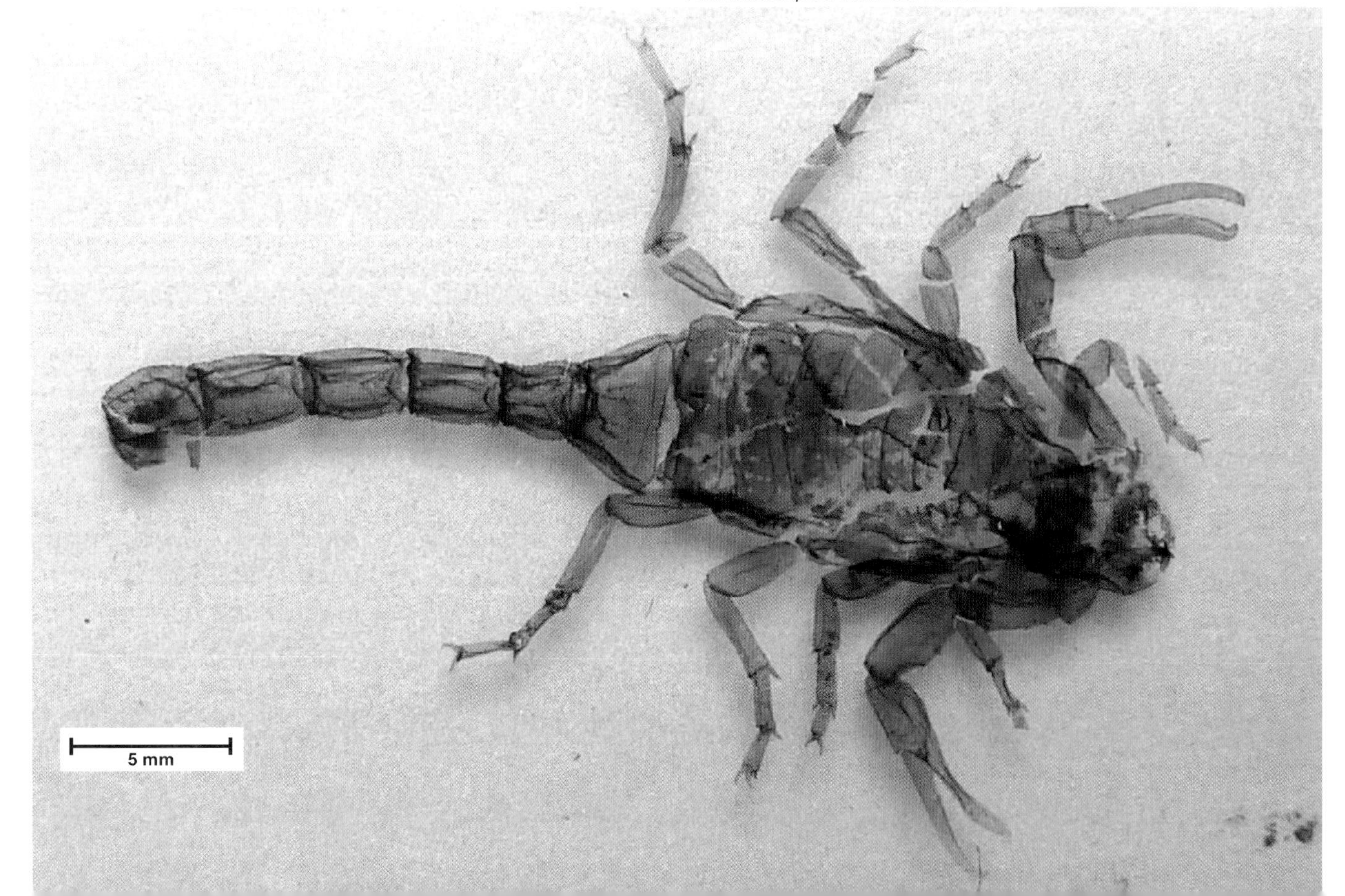

66

Archidesmus / Millipede / Devonian

Phylum	**Arthropoda**
Class	**Myriapoda, Diplopoda**
Gen et sp	***Archidesmus macnicolli***
Locality	**Tillywhandland Quarry, Forfar, Angus**
Age	**Early Devonian**
Stratigraphy	**Dundee Formation, Arbuthnott Group, Lower Old Red Sandstone**

Millipedes are a very ancient group of terrestrial animals – in fact *Pneumodesmus newmani*, a millipede from the Cowie Formation at Stonehaven, is mid- to late Silurian in age and has openings called spiracles on the segments, which prove that it was an air-breathing animal (Wilson and Anderson 2004). This is the oldest known air-breathing animal.

Archidesmus is younger, from the early Devonian, and also lived on land amongst some of the earliest land plants. The specimen was found in association with fish fossils such as *Ischnacanthus*, and plant fragments such as *Parka*. The animal is 25mm long and there are about 28 segments with two pairs of thin legs to each body segment – apart from body segment eight, where the front pair of legs is modified and greatly swollen. This modified pair of legs may have acted as claspers during mating. The millipede and plants were transported in rivers and lakes to their final resting place. It seems remarkable that such an animal could be transported in this way and remain complete, including the delicate legs.

Since millipedes similar in structure are still alive today, we can solve the problem. When a millipede falls into water it floats on one side, thus it can still breathe through spiracles on the uppermost side of the body. It can survive like this for days, and thus can drift a long way in a river. When it dies it still takes a long time to disintegrate, and it actually breaks into sections before the legs fall off. Thus it is easy to explain the presence of a terrestrial millipede in sediments deposited in a deep lake.

66 *Archidesmus*, a millipede from the Lower Old Red Sandstone.

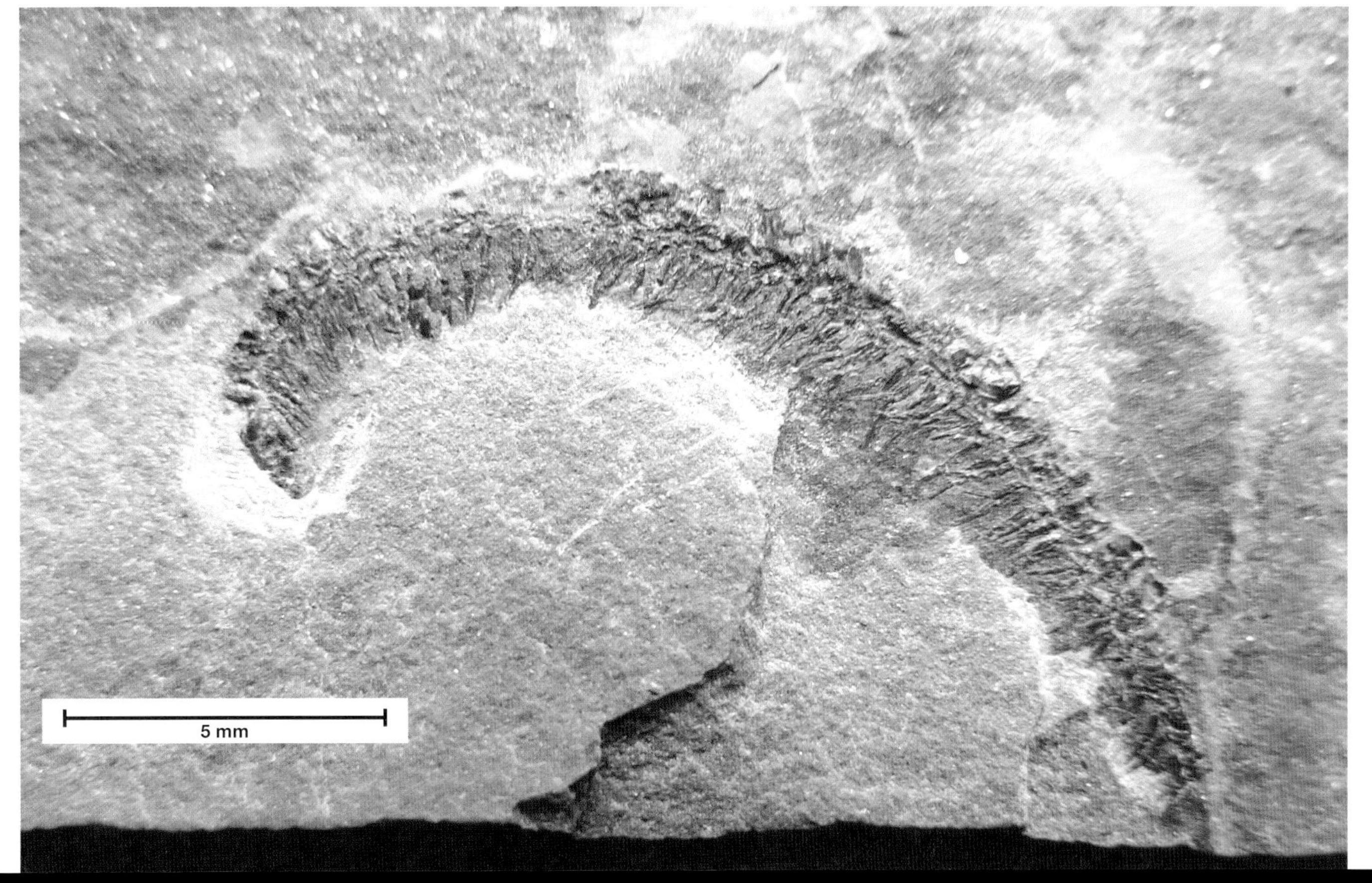

67

Palaeocharinus / Trigonotarbid / Devonian

Phylum	**Arthropoda**
Class	**Chelicerata**
Gen et sp	***Palaeocharinus rhyniensis***
Locality	**Rhynie, Aberdeenshire**
Age	**Early Devonian**
Stratigraphy	**Rhynie Cherts Unit, Dryden Flags Fm., Lower Old Red Sandstone**

Palaeocharinus belongs to the group of arthropods that includes horseshoe crabs, scorpions, spiders and mites. It is similar to a spider, but lacks spinnerets to produce silk, and has a distinctly segmented abdomen. It belongs to an extinct group called trigonotarbids. 'Trigs' are the most common arthropods in the Rhynie chert, and are found both as complete animals, and moulted parts. The preservation is occasionally so good that book lungs are preserved, proving that this was a terrestrial animal. *Palaeocharinus* had large fangs, and was clearly a predator of other small arthropods. It lived amongst living plants and plant litter, and specimens have been found fossilised within empty plant sporangia and hollow plant stems. Such sheltered places would have been safe refuges for moulting, and places to deposit eggs. Other specimens have been found clinging to upright plant stems, as in the example shown here. In the Early Devonian there were few plants more than 50cm high, and *Palaeocharinus* would have hunted in this mini forest in the company of centipedes, millipedes, mites, harvestman spiders and primitive insects. A recent addition to the Rhynie fauna is a new species, *Palaeocharinus tuberculatus*, described by Fayers *et al.* (2005) from the Windyfield chert, and illustrated with a reconstruction by Stephen Caine. The Windyfield chert was discovered in 1988 near the Rhynie chert site, and has produced several new arthropods (Anderson and Trewin 2003), making the Rhynie arthropod fauna the most diverse Early Devonian terrestrial and freshwater fauna anywhere in the world.

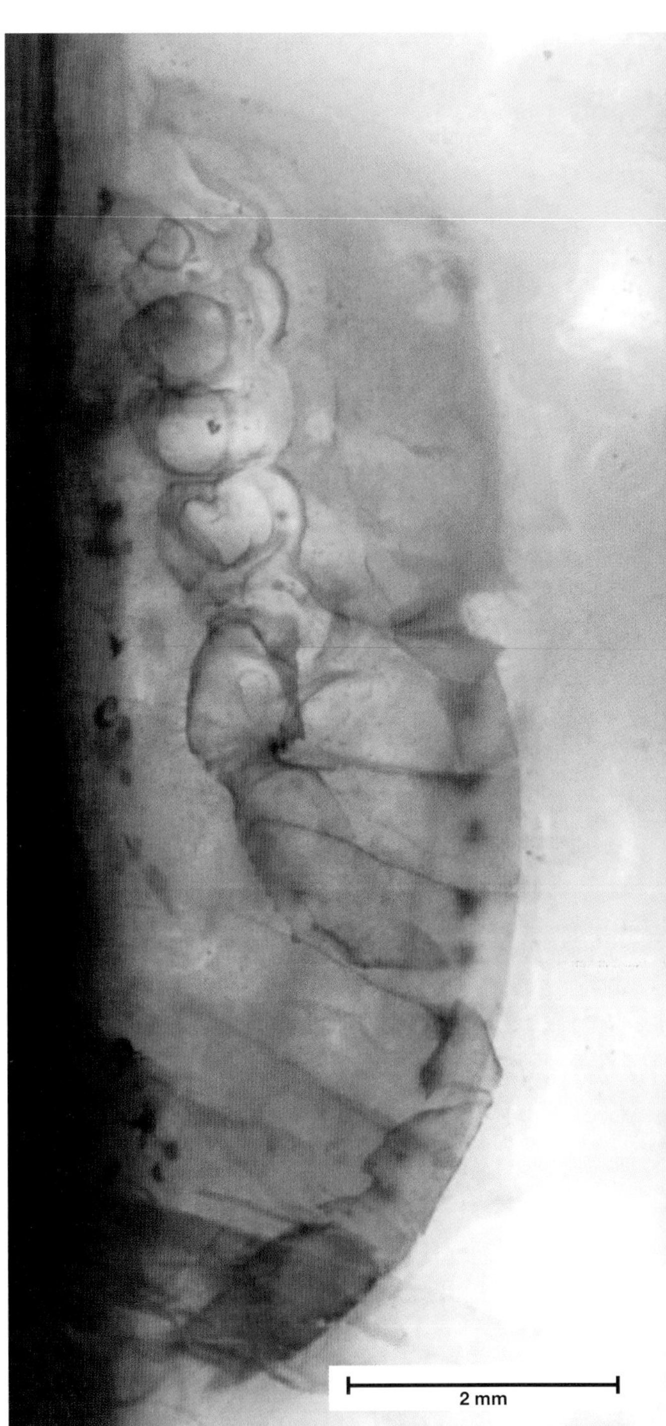

67a *Palaeocharinus* preserved clinging to a plant stem in the Rhynie chert.

67b Model of *Palaeocharinus*.

68

Tealliocaris / Crustacean / Carboniferous

Phylum	**Arthropoda**
Class	**Crustacea**
Gen et sp	***Tealliocaris woodwardi***
Locality	**Cheese Bay, near Gullane, East Lothian**
Age	**Early Carboniferous**
Stratigraphy	**Cheese Bay Shrimp Bed, Oil-Shale Group**

This superb shrimp occurs in abundance in the Cheese Bay Shrimp bed at Cheese Bay near Gullane, within the Oil Shale Group of the Early Carboniferous. The shrimp bed is a hard, finely laminated rock with laminae of dolomite and organic matter, which was deposited as fine mud in a large lake or lagoon within a delta swamp. The lamination is probably of seasonal origin. Periodically the sea broke into the lagoon, bringing a few marine animals such as orthocone cephalopods, and possibly killing off the shrimps by the change in salinity (Hesselbo and Trewin 1984). Today, shrimps are also common in lagoonal environments, and can occur in enormous numbers. The ancient lagoon was rich in organic matter, probably in the form of algae and micro-organisms that were food for the shrimps. At times organic matter accumulated to form the oil-shale that was exploited in the Lothians, leaving a legacy of 'bings' of red burnt shale from which oil had been distilled.

Many of the shrimps are preserved in white phosphate on dark limestone, and details of delicate appendages such as the antennae are clearly visible. Using specimens crushed in a variety of positions, Briggs and Clarkson (1985) were able to reconstruct the detailed anatomy of the animal. There are several other shrimp beds within the Oil Shale Group that have been described by Cater *et al.* (1989).

68 The Carboniferous shrimp *Tealliocaris* from the Cheese Bay shrimp bed.

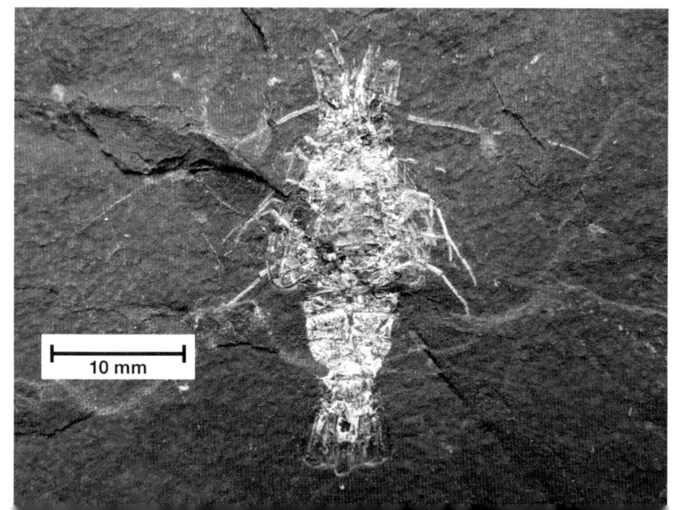

69

Palaemysis / Crustacean / Carboniferous

Phylum	**Arthropoda**
Class	**Crustacea**
Gen et sp	***Palaemysis dunlopi***
Locality	**Bearsden, Glasgow**
Age	**Carboniferous, Namurian**
Stratigraphy	**Manse Burn Formation**

This shrimp was poorly known from isolated parts until complete specimens were found by Stan Wood in the 'Bearsden Shark Dig' in 1981 (see *Akmonistion*). The excavation produced a superb fauna first reported by Wood (1982), and also greatly increased interest in palaeontology in Bearsden and beyond. This shrimp has been described by Clark (1991) and appears to have been a predator; it has one pair of antennae modified into structures with sharp spines on the inner sides. These look as though they would have been very effective in grasping and holding prey. With an overall length of 8cm it is larger than other shrimps such as *Crangopsis*, with which it is associated in a 'shrimp bed'. The shrimps were found at several levels in the excavation and appear to have lived in brackish water conditions, maybe in a lagoon during periods when the area was cut off from open marine seas. There were rapid variations in sea level during the Carboniferous, and periods of low sea level may have been responsible for changing the environment at Bearsden from open sea with a marine fauna to lagoonal conditions with reduced salinity. Deposition in fine mud of a low energy lagoonal environment without strong waves or currents would have favoured the preservation of the delicate shrimps.

69 *Palaemysis*, a predatory Carboniferous shrimp from Bearsden.

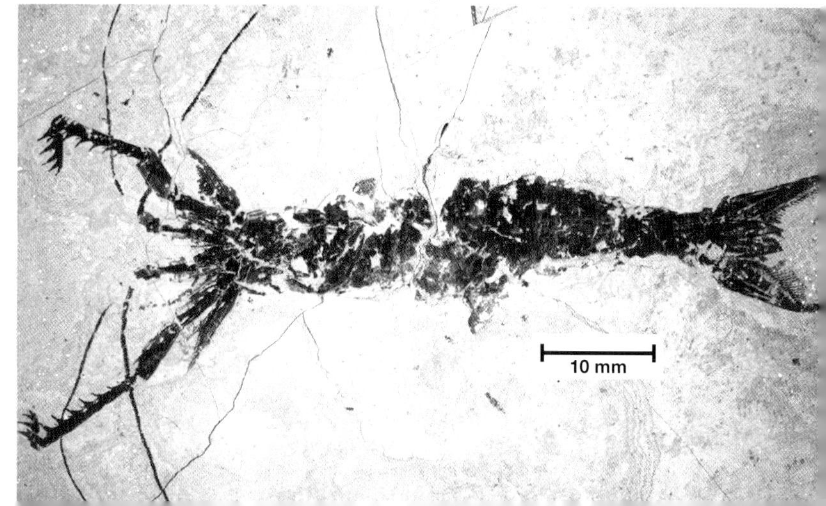

70

Estheria / Crustacean / Jurassic

Phylum	**Arthropoda**
Class	**Crustacea, Branchiopoda**
Gen et sp	***Estheria***
Locality	**Elgol, Isle of Skye**
Age	**Mid-Jurassic**
Stratigraphy	**Lealt Formation, Great Estuarine Group**

'*Estheria*' is a general name applied to the shells of small pod-shrimps, of which vast numbers inhabited muddy lagoons on a delta top adjacent to the Jurassic sea in Skye. The animal is a branchiopod crustacean, and many modern representatives of the group live in semi-arid areas where temporary pools are created by infrequent rains. Eggs that may have lain dry and dormant in the mud for many years hatch when wetted. The animals grow rapidly and produce eggs within a few weeks. When the pool dries out, the eggs remain in the sediment ready to start another generation after the next rains. There are a number of species of conchostracans present in the shales of the Lealt Formation, and these have been described by Chen and Hudson (1991), who identified 12 species in 7 genera, and showed the ranges of these species within the Great Estuarine Group. Large numbers of ostracods are also present; they are also crustaceans, and are commonly associated with branchiopod crustaceans in lagoonal Jurassic sediments.

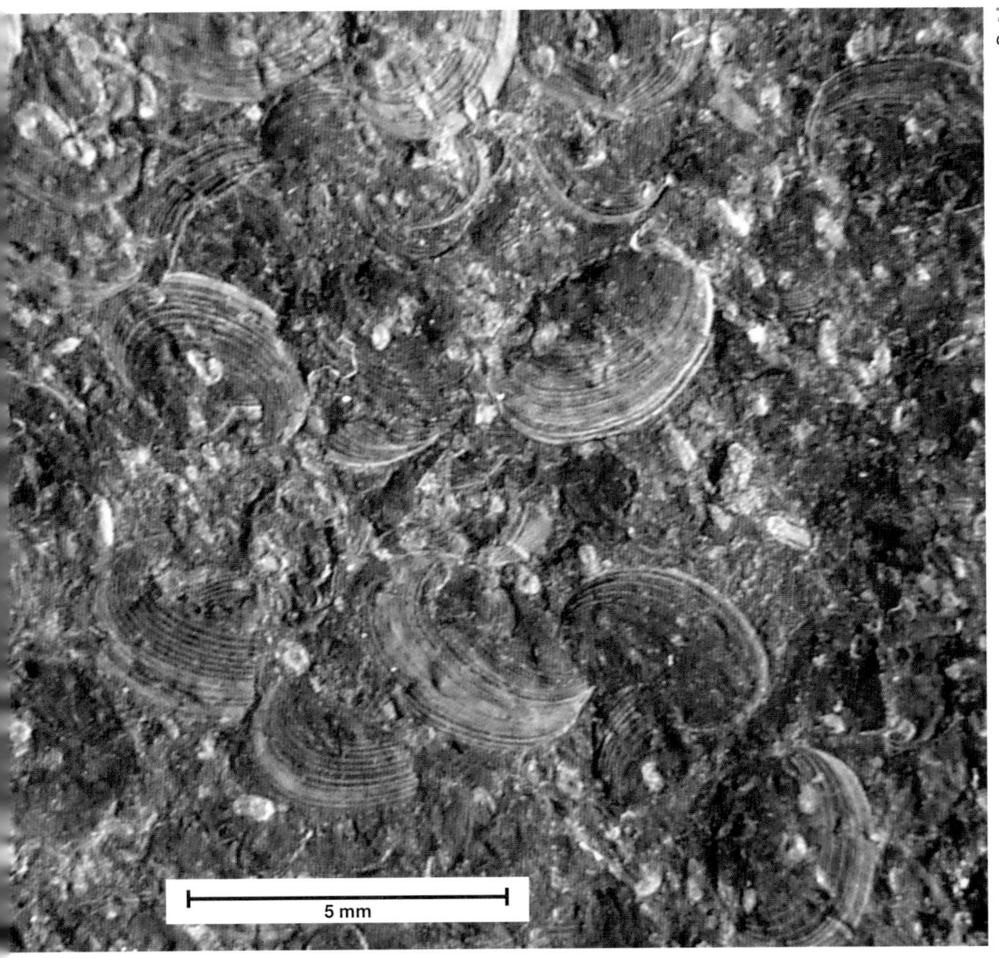

70a The bivalved crustacean *Estheria* and tiny ostracods crowded in lagoonal Jurassic shales from Skye.

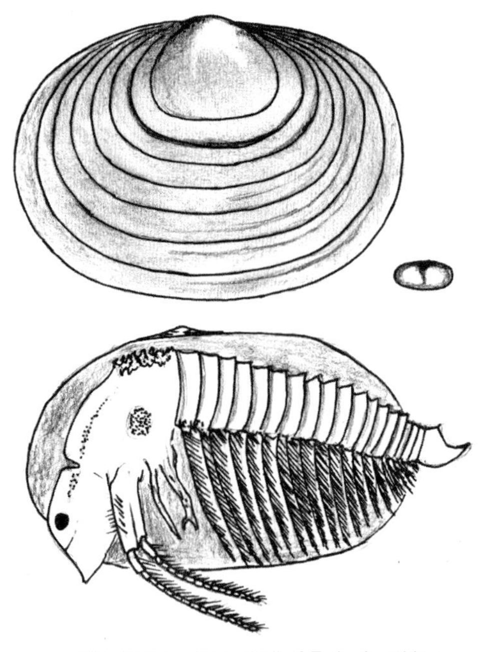

70b Sketch of the shell of *Estheria*, with a small bean-shaped ostracod at the same scale, and a modern crustacean similar to *Estheria* within a shell.

71

Idoptilus / Insect / Carboniferous

Phylum	**Arthropoda**
Class	**Insecta, Paleodictyoptera**
Gen et sp	***Idoptilus peachii***
Locality	**Greenhill Quarry, Kilmaurs, Ayrshire**
Age	**Late Carboniferous, Westphalian**
Stratigraphy	**Middle Coal Measures**

Insect fossils are generally very rare because most are delicate and have low preservation potential. Those insects that pass part of their life cycle in water stand a better chance of preservation, and *Idoptilus* was such an animal. Although originally interpreted as a cockroach nymph, in 2010 it was reinterpreted by Andrew Ross of National Museums Scotland as the nymph stage of an extinct type of dragonfly, found only in the Carboniferous and Permian. The nymph had a small head set between two lobes. Four wing pads show curving veins that are also characteristic of the wings of the mature flying insect. Some of the dragonflies of the Carboniferous were very large, with wingspans of up to 50cm. The atmosphere in the Carboniferous was richer in oxygen than our present atmosphere, allowing insects to grow larger as more oxygen could enter the body.

Idoptilus is one of only five Carboniferous insects known from Scotland. It is preserved in a small ironstone nodule that formed around the dead nymph. It was found in shale overlying a coal seam, and would have lived in the Carboniferous coal forests where there was plenty of water in which to lay eggs and food for the young nymphs. When fully grown the nymph would have climbed out of the water and moulted into the mature dragonfly stage, just as modern dragonflies do today.

The specimen resides in the collection of the Dick Institute in Kilmarnock, close to where it was found. It is part of the Ben Peach Collection, most of which is in the British Geological Survey collections, since Ben Peach was a fossil collector for the Survey. The specimen was first described in 1887 and named in honour of the finder by H. Woodward.

71 A rare Carboniferous insect, *Idoptilus*, in a small nodule.

VERTEBRATES

Vertebrates, animals with backbones, are also known as craniates due to features of the head possessed by all members. Not all animals classed as vertebrates actually have vertebrae, and not all have bone. The origin of vertebrates within the phylum Chordata has been a matter for speculation and divided opinion for many years, but the closest relatives of vertebrates are generally considered to be the sea squirts and amphioxus. In truth, the boundaries are blurred. There are several morphological features, such as posession of a notochord (a stiff fluid-filled rod) and myotomes (v-shaped muscle blocks) shared by chordates and vertebrates, but it is a matter of opinion where a line should, or even can, be drawn. As one might expect, evolution has provided a continuum, and it is Man's desire to make hard classifications in this continuum that is the problem. In simple terms, vertebrates can be divided into fish, amphibians, reptiles and mammals. However, these are not hard and fast divisions into which every vertebrate can be placed.

The conodont animal *Clydagnathus* was discovered in 1983, and we can now be sure that this animal fits within our broad definition of 'fish'. Many early agnathan fish related to lampreys lack hard parts, and thus are seldom well preserved, and the fossil record is poor. However, fish with bone have a much more detailed geological record. Thus preservation is an important factor.

Environment also plays a part; fish are relatively common as fossils since they live in water and can be easily buried in sediment. On the other hand, terrestrial animals generally have to be washed into rivers, lakes or the sea to be preserved. On a land surface a carcass is rapidly destroyed by scavengers and organic material is oxidised, and no potential fossil remains. An exception is the situation where a sand dune covers a dried carcass in a desert environment. Hence the geological record of terrestrial animals is worse than the record of aquatic animals.

Just as it is difficult to identify the earliest vertebrate, it is increasingly difficult to define the transitions from fish to amphibian, amphibian to reptile, and reptile to mammal. Again, it needs to be stressed that in recent years fossils such as *Elginerpeton* have been found that show stages in the evolution of fish to amphibian, and *Westlothiana* shows features of the transition from amphibian to reptile. These recent finds in Scotland are filling the gaps in the evolutionary record of the vertebrates. Creationists please note, the 'missing links' on which you pin false hope are increasingly to be found in museum drawers! I am always impressed by how much of the biological record has been preserved, considering the chain of events required to produce a fossil of any kind.

In the sections below on fish, amphibians, reptiles and mammals, a few of the important and interesting Scottish finds are described. Scotland has an extraordinary wealth of world-famous vertebrate material, ranging from the Devonian fishes of the Old Red Sandstone, to the Carboniferous amphibians of East Kirkton, and the Permo-Triassic reptiles of Elgin. What we lack are complete skeletons of large dinosaurs; maybe one day we will find one, probably on the Isle of Skye.

Vertebrates – Fish

Scotland has a magnificent series of fossil fish localities that yield particularly well-preserved fossil fish. These localities have been important in the history of the study of fossil fish, particularly through the early efforts of Hugh Miller on the Old Red Sandstone, and the monograph by Agassiz (1844).

Particularly important to science are the faunas of primitive agnathan fish such as *Jamoytius, Birkenia, Lasanius* and *Thelodus* from the Silurian of the Lesmahagow area. From the Early Devonian Lower ORS of the Forfar area there are the superb articulated acanthodians such as *Climatius, Ischnacanthus Parexus* and *Mesacanthus* that inhabited rivers and lakes of the Midland Valley. The Middle ORS flagstones of Caithness and Orkney contain numerous fish beds, of which the Achanarras fish bed yields some 16 genera of fish preserved in the deep lake laminites of Lake Orcadie. The preservation of the fish is again exceptional, with some impressions of soft tissue in agnathans, as well as fine specimens of arthrodires and crossopterygians. In the nineteenth century, surfaces covered in *Holoptychius* were discovered from the Upper ORS of Dura Den in Fife, now found distributed in many museum collections. The fish died when pools dried out and carcasses were covered with wind-blown sand. In the Carboniferous the highlights are the shark fauna excavated by Stan Wood at Bearsden, and the fish preserved in nodules at Wardie Shore.

72

Clydagnathus / Fish, Conodont animal / Carboniferous

Subphylum	**Vertebrata**
Class	**Agnatha, Conodonta**
Gen et sp	***Clydagnathus cavusformis***
Locality	**Granton, Edinburgh**
Age	**Lower Carboniferous**
Stratigraphy	**West Lothian Oil Shale Fm., Granton Shrimp Bed**

Conodonts are small (most about a millimetre in size) phosphatic toothlike fossils that have been known and described since 1856. There had been much speculation as to the type of animal they came from, but it was clear that the animal was bilaterally symmetrical and that several types of conodonts were present in a single animal. This was known from conodont assemblages seen in some shales, but no body was preserved. Conodonts are found in marine rocks from the Late Cambrian to the Triassic, and are used to determine stratigraphic age, since they evolved rapidly and have distinctive shapes. However, it was not until 1983 that the first fossil of an animal containing conodonts was described (Briggs *et al.* 1983). Ten examples have been found in the Granton Shrimp Bed at Edinburgh, ranging from 21 to 55mm in length and preserving impressions of soft parts. They show an eel-like laterally flattened shape with large eyes, V-shaped muscle blocks, a notocord and a tail with fin-rays. The conodonts are grouped in the throat of the animal, and were probably used to trap and pass prey to the stomach.

Clydagnathus settled arguments regarding the affinity of conodonts, and it is certainly one of the most important Scottish fossil finds of all time. Conodont animals are now also known from the Silurian of Wisconsin, USA and the Ordovician of South Africa. As with buses or Edinburgh trams, you wait nearly 150 years and three turn up!

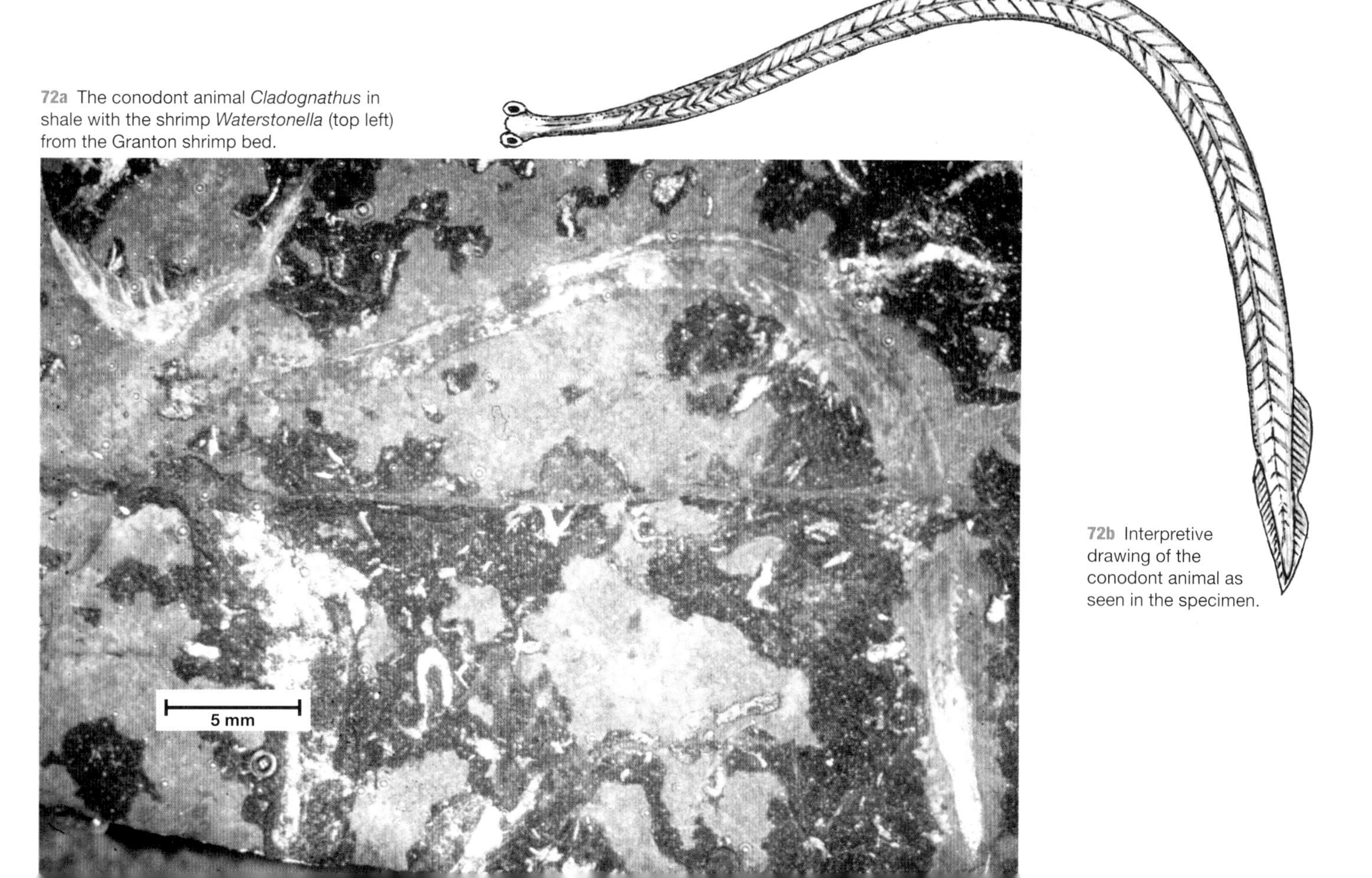

72a The conodont animal *Cladognathus* in shale with the shrimp *Waterstonella* (top left) from the Granton shrimp bed.

72b Interpretive drawing of the conodont animal as seen in the specimen.

73

Jamoytius / Fish / Silurian

Subphylum	**Vertebrata**
Class	**Agnatha, Anaspida**
Order	**Jamoytiidae**
Gen et sp	***Jamoytius kerwoodi***
Locality	**Logan Water, Lesmahagow**
Age	**Silurian**
Stratigraphy	**Patrick Burn Formation, Priesthill Group**

Jamoytius is a jawless fish that is related to the modern lamprey. It had a sucker mouth and a row of gill openings. There were eyes and a distinctive downturned tail fin. Internally it posessed v-shaped muscle blocks and a stiff rod, or notocord, rather than a backbone of vertebrae. This very primitive fish is rare, and is associated with other similar jawless fishes, the whole assemblage being important for the study of vertebrate evolution. The famous fossil sites in the Lesmahagow area were discovered in the nineteenth century, and in 1899 amateur palaeontologists from the Glasgow Geological Society led by Peter McNair set up annual summer camps, called 'Camp Siluria' to exploit the beds, even using gunpowder to blast the beds. Alex Ritchie studied *Jamoytius* in detail, noting the importance of this fossil in the development of fish (Ritchie 1968). The sites are now protected, and quarrying is not permitted. A modern 'Camp Siluria' might well yield new material relevant to the evolution of fish.

73a The primitive agnathan fish *Jamoytius* showing v-shaped muscle blocks.

73b Reconstruction of *Jamoytius*.

74

Birkenia / Fish / Silurian

Subphylum	**Vertebrata**
Class	**Cephalaspidomorphi, Anaspida**
Order	**Birkeniida**
Gen et sp	***Birkenia elegans***
Locality	**Birkenhead Burn, near Lesmahagow**
Age	**Silurian**
Stratigraphy	**Fish Bed Formation**

Birkenia is an armoured jawless fish with well-developed elongate bony scales and distinctive modified hook-like scales along the dorsal surface. It has the typical downturned tail of an anaspid fish. This fish occurs in association with other jawless fishes (*Lasanius, Ateleaspis, Lanarkia* and *Logania*) and eurypterids that lived in a lake or lagoon. When these fish were first found in the area by James Young in 1896 they were the oldest complete fish known in the world. They were described by Traquair (1840–1912) who was Keeper of Natural History at the Royal Scottish Museum. However, he made an error in interpretation, and thinking that they should have all had heterocercal tails (like sharks), he reconstructed some upside-down. Lower down in the succession marine fossils such as brachiopods and trilobites are found, but marine conditions gradually changed to freshwater as the basin became isolated from the ocean during the final stages of the closure of the Iapetus Ocean. The fish and arthropods evolved and survived this major change in environment, but the marine faunas died out.

This specimen was collected by Professor F.H. Stewart, late of Edinburgh University, who formed a fine collection of fossil fish that is now in the National Museums, Scotland collection in Edinburgh. His main scientific work was connected with the salt industry in Yorkshire through ICI. He acted as scientific advisor to Government and he was honoured with a knighthood. His wife is the well-known author of best-selling romantic novels, Mary Stewart.

74a *Birkenia*, an agnathan with long thin scales and dorsal spines.

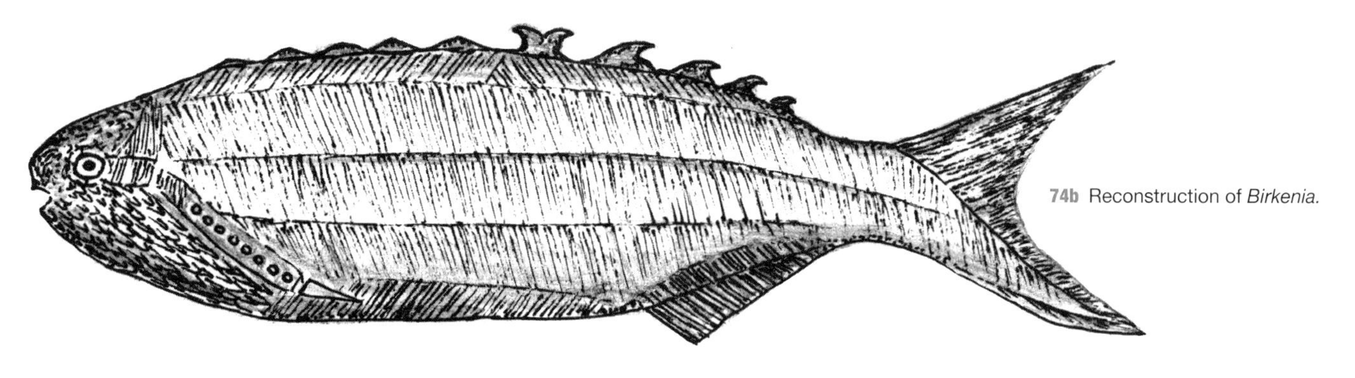

74b Reconstruction of *Birkenia*.

75

Zenaspis (*Cephalaspis*) / Fish / Devonian

Subphylum	**Vertebrata**
Class	**Cephalaspidomorphi**
Order	**Cephalaspidida**
Gen et sp	***Xenaspis pagei***
Locality	**Tillywhandland Quarry, near Forfar, Angus**
Age	**Early Devonian**
Stratigraphy	**Dundee Formation, Arbuthnott Group, Lower Old Red Sandstone**

Zenaspis is a relatively new name for a fish previously included in the genus *Cephalaspis*. As studies continued, it was recognised that fish known as *Cephalaspis* included a great variety of fish with features so different that they merited separate generic names. Hence, following scientific revision, this fish is now known as *Zenaspis*.

The cephalaspid fishes are agnathans, meaning that they do not have jaws. The characteristic crescentic headshield is the part most frequently found as a fossil; complete specimens are very rare. The mouth was on the underside of the headshield, and on the top are two closely spaced eyes and a nasal opening. Sensory fields, where the bone is rich in nerve canals, occur near the dorsal margin, and may have been electric organs. The tail was rather weak, and this fish was probably a poor swimmer, spending its life grubbing for food on the bottom of lakes and rivers. One reason why headshields are the most commonly found part of the fish is that the distinctive shape catches the eye; the elongate body scales are rarely noticed. The headshield was also the strongest part of the skeleton, so is more likely to have survived.

James Powrie (1814–1895) who owned the Reswallie Estate, formed an important collection of fossil fish and arthropods from the Forfar area and published about 15 papers on the geology and fossils of the area. His pioneering contribution to geology in the Forfar area has been discussed by Davidson and Newman (2003). He started collaboration with E. Ray Lankester on *A Monograph of the Fishes of the Old Red Sandstone of Britain*. Part 1 of this work on the Cephalaspids was authored by Lankester (1868) who named one species after Powrie, and others after Page, Murchison and Agassiz. However, he did not contribute directly to the monograph, and Part 2 of the work was authored by Traquair (1894–1914).

A classic work on cephalaspid fish is the monograph by Stensio (1932), in which Powrie's material is described in much more detail. A valuable recent contribution on the phylogeny of these fishes is the detailed study of Sansom (2009). Thus Powrie's collection, now in National Museums Scotland, continues to be studied.

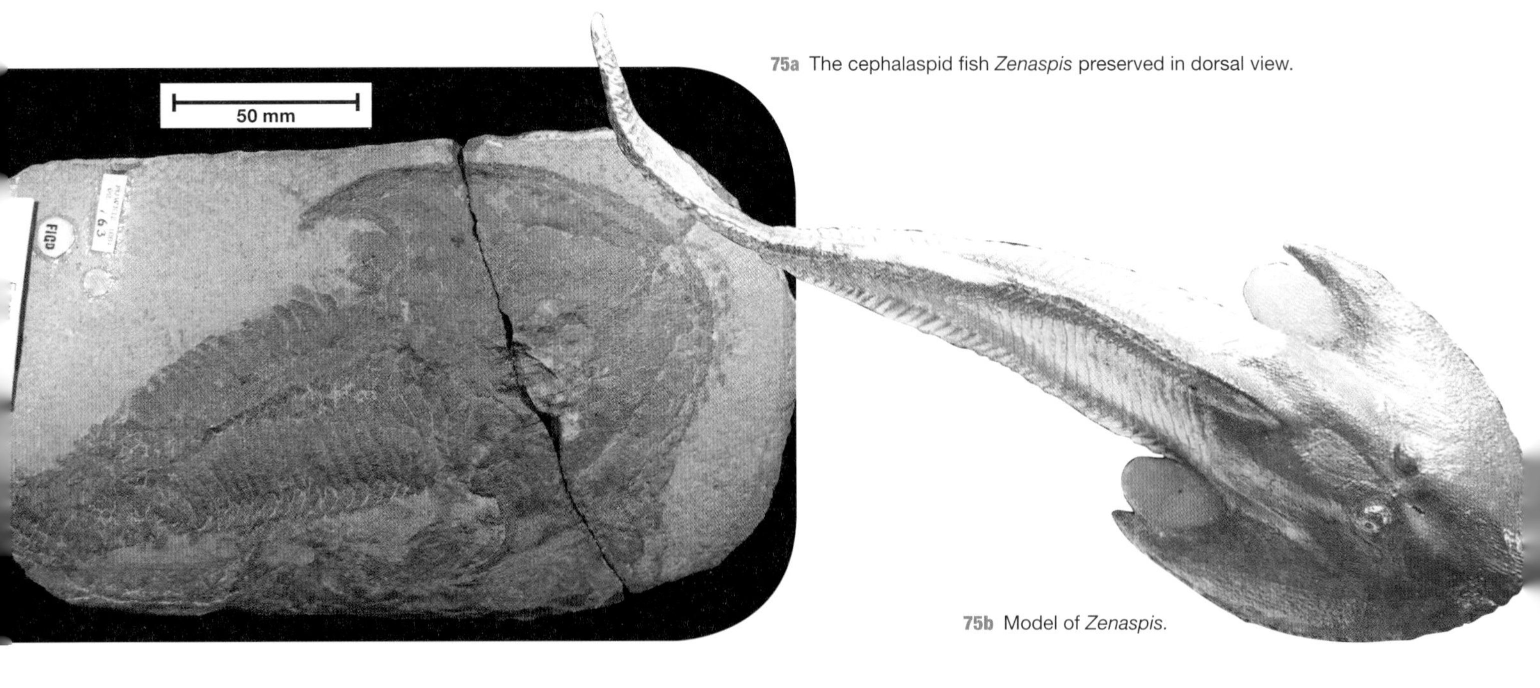

75a The cephalaspid fish *Zenaspis* preserved in dorsal view.

75b Model of *Zenaspis*.

76

Ischnacanthus / Fish / Devonian

Subphylum	**Vertebrata**
Class	**Acanthodii**
Gen et sp	***Ischnacanthus gracilis***
Locality	**Tillywhandland, near Forfar, Angus**
Age	**Early Devonian**
Stratigraphy	**Dundee Formation, Arbuthnott Group, Lower Old Red Sandstone**

Ischnacanthus is a representative of the acanthodian, or 'spiny-finned' fishes from the Lower Old Red Sandstone of Strathmore. These fish were the earliest jawed fishes. They occur in Silurian rocks, and invaded fresh water early in the Devonian. They rapidly diversified into a group with a variety of body forms, characterised by differing numbers and types of spines. They occupied a range of environments in lakes and rivers of the Old Red Sandstone continent. At Tillywhandland the fish occur in laminated shales overlying a thick sandstone that was quarried for building stone. Trewin and Davidson (1996) suggested that following the deposition of the sandstone by rivers, the area was rapidly flooded to produce a deep lake. Volcanic activity may have blocked the river system to produce the lake.

The quarries in the Forfar to Dundee region produced many fine specimens in the nineteenth century for James Powrie's collection. At the time the quarries were working the quarrymen would save fossils for Powrie, but nowadays the quarries are overgrown, and it is not easy to find fossil fish. Powrie is the author of the genus *Ishnacanthus* (Powrie 1864), and the superb specimen of *Ishnacanthus* illustrated is a modern find that has been carefully prepared from a carbonate nodule using weak acid. The Strathmore area is one of the most celebrated sites in the world for preservation of complete Devonian acanthodians.

The spines supported the leading edges of fins, and also served as defence. This fish has two dorsal fin spines, an anal fin spine and paired pectoral spines. It had well-developed jaws with teeth, a streamlined body, and was clearly a predator. Coprolites (fish excreta) are found in the fish bed, and contain the spines of the smaller acanthodian *Mesacanthus*, and even small *Ishnacanthus*. The prey was swallowed whole, and head-first. This can be deduced from the orientation of finspines in the coprolites. The acanthodian fishes have many shark-like features, and it is thought that sharks may have evolved from an acanthodian ancestor.

76a A fine acid-prepared specimen of the predatory acanthodian *Ischnacanthus* from the Lower Old Red Sandstone.

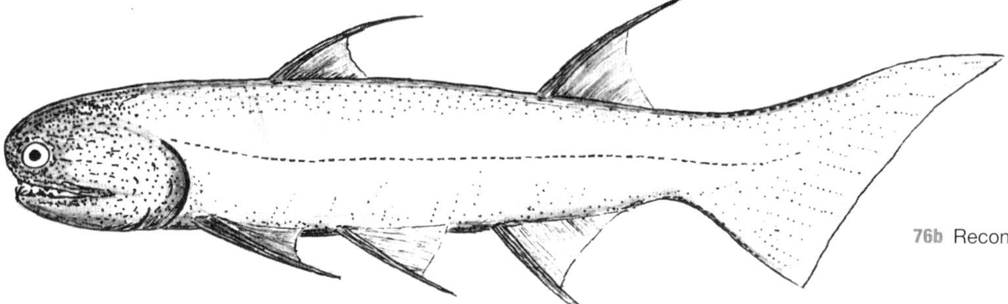

76b Reconstruction of *Ischnacanthus*.

77

Climatius / Fish / Devonian

Subphylum	**Vertebrata**
Class	**Acanthodii**
Gen et sp	***Climatius reticularis***
Locality	**Tillywhandland, near Forfar, Angus**
Age	**Early Devonian**
Stratigraphy	**Dundee Formation, Arbuthnott Group, Lower Old Red Sandstone**

This acanthodian fish with stout ornamented spines is rarely found complete; hence the illustrated specimen is exceptional. It has been prepared from a limestone nodule using weak acid and shows the details of the spines very clearly. The pectoral spines were fused to the pectoral girdle, giving the fish a flat belly, probably indicating a bottom-feeding habit. The jaws were weak and seem to lack teeth, supporting a scavenging mode of life. The fish is partly preserved in ventral view and shows the series of paired spines on the belly. The body is covered with tiny scales and the head is composed of many small plates. The head partly decayed prior to preservation – this is common in fish fossils, and due to decay starting, where bacteria and scavengers can attack the soft parts of a carcass through the mouth and gill openings. Acanthodian fishes of the World are summarised by Denison (1979) in the *Handbook of Palaeoichthyology*, where the great importance of the Scottish faunas is readily apparent.

77a The acanthodian *Climatius* in a calcareous nodule from the Lower Old Red Sandstone.

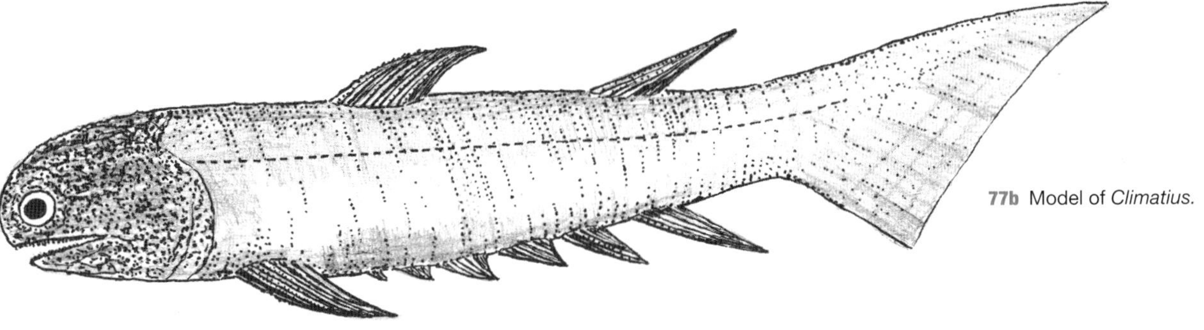

77b Model of *Climatius*.

78

Coccosteus/ Fish / Devonian

Subphylum	**Vertebrata**
Class	**Placodermi**
Order	**Arthrodira**
Gen et sp	***Coccosteus cuspidatus***
Locality	**Achanarras Quarry, Caithness**
Age	**Mid-Devonian**
Stratigraphy	**Caithness Flagstone Group, Middle Old Red Sandstone**

Coccosteus is an armoured arthrodire, and belongs to an extinct group of fish, the placoderms. Hugh Miller had problems reconstructing this fish and *Pterichthyodes*. Both have bony plates with a similar ornament, and until he found reasonably complete specimens he thought he was dealing with a single animal.

The head and part of the body are covered with strong bony armour, and there is a gap in the armour between the skull and body to allow the fish to open its mouth with a wide gape. The rear half of the fish was naked, and it is usually preserved so that the vertebral column can be seen. The vertebral column does not have bony centra; only the neural and haemal arches are ossified. The dorsal fin and long thin tail are usually preserved, but other fins are rarely seen. The jaws of *Coccosteus* lack true teeth, but were equipped with strong bony plates with sharp shearing edges that fitted perfectly, and would have supplied a bite that could have removed lumps of flesh. This fish was certainly a predator, since a few specimens have been found with stomach contents containing bones of *Dipterus* and acanthodians, together with small pebbles used to grind up the food.

Virtually all the specimens of *Coccosteus* found in the Devonian of northern Scotland are of adult fish about 30cm long. It is interesting to ask why we find no juvenile specimens. Maybe the fish did not breed in the Orcadian Lake, but migrated there when adult. It has been suggested that the fish may have spawned and spent their youth in rivers that fed the lake, or they may even have migrated to the lake from the sea when adult (Trewin 1986). There are several different closely related arthrodires that seem to represent an evolving stock that lived in the Orcadian Lake; thus different levels of the Caithness flagstones can be recognised on the basis of these fish fossils (Miles and Westoll 1963).

Arthrodires ruled the seas of the Devonian, and *Dunkleosteus*, a giant fish over 20m long, was the largest vertebrate of the Devonian, which, with fearsome jaws, could have swallowed any other fish in the sea.

078a The predatory arthrodire *Coccosteus* from the Middle Old Red Sandstone of Achanarras Quarry, Caithness.

078b Model of *Coccosteus*.

078c Looking for fish in the tips at Achanarras Quarry, Caithness.

79

Pterichthyodes / Fish / Devonian

Subphylum	**Vertebrata**
Class	**Placodermi**
Order	**Antiarchi**
Gen et sp	***Pterichthyodes milleri***
Locality	**Achanarras Quarry, Caithness**
Age	**Mid-Devonian**
Stratigraphy	**Caithness Flagstone Group, Middle Old Red Sandstone**

This bizzare armoured fish was first described by Hugh Miller (1841) in his famous book *The Old Red Sandstone*. He pieced together most of the features of the fish from specimens he found on the beach at his home town of Cromarty. The cottage in Cromarty where Miller was born, and the adjacent family home, now the Hugh Miller Museum, are National Trust for Scotland properties. Both are open to visitors and contain displays relating to Miller's life, including geology displays with some of the fossils he collected.

When Miller first found this fish he was not even sure that it was a fish, only being convinced when he discovered the scaly tail with a dorsal fin. The fish was scientifically described by Agassiz (1844), who gave it the specific name *milleri* in honour of the finder.

The fish has a box-like armoured body from which protrudes a scaled body with heterocercal tail and a dorsal fin supported by a spine. The most remarkable features are the armoured pectoral fins that look like paddles and articulate with the body by means of ball and socket joints. There are forms with wide and narrow pectoral paddles, thought to be males and females, but we cannot tell which was which. The mouth is on the flat underside and had very weak jaws, and the eyes are on the dorsal surface gazing forwards and upwards. All these features suggest that *Pterichthyodes* was a weak swimmer, and a bottom dweller, sifting the sand and mud for particles of food.

The specimens found at Achanarras Quarry were deposited in deep water after carcasses had drifted from shallow water areas out into the lake before sinking. Some fossils show that carcasses had started to decay and fall apart before they were deposited on the lake floor (Trewin 1986). This fish could not have lived in the deep water of the lake because there was little or no oxygen present. A full size range of individuals from juveniles only a centimetre long to adults some 30cm long are found, implying that this fish passed its full life history in the lake. This is in sharp contrast with the size distribution of *Coccosteus* as described above.

79a *Pterichthyodes*, a large specimen showing typical preservation with outstretched pectoral paddles and detached headshield.

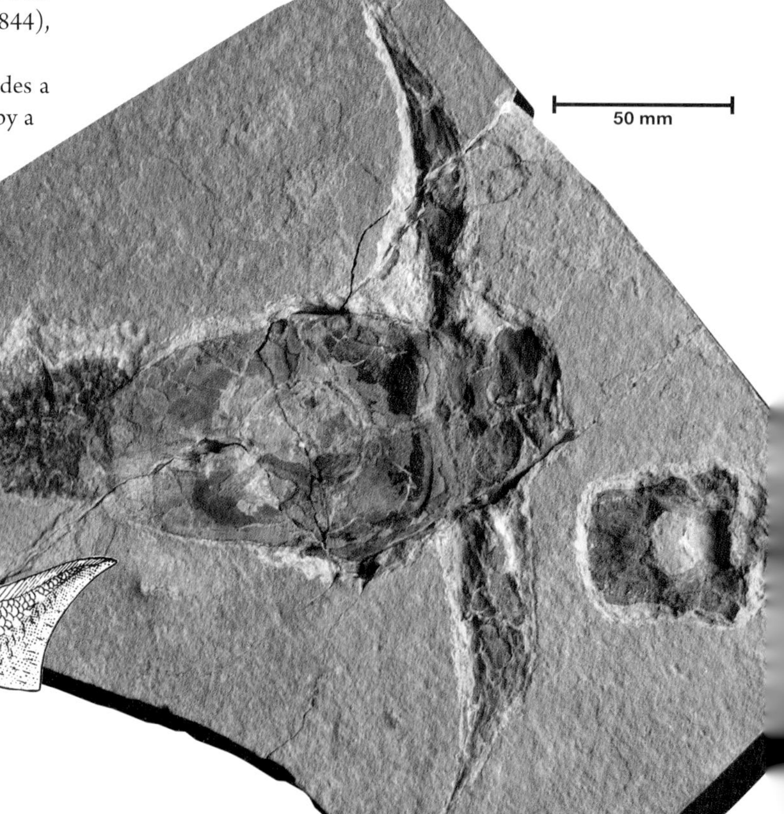

79b Reconstruction of *Pterichthyodes*.

80

Palaeospondylus / Fish / Devonian

Subphylum	Vertebrata
Class	uncertain
Gen et sp	***Palaeospondylus gunni***
Locality	Achanarras Quarry, Caithness
Age	Mid-Devonian
Stratigraphy	Caithness Flagstone Group, Middle Old Red Sandstone

This strange little tadpole-shaped fish only a few centimetres long was described by Traquair (1890) and has puzzled palaeontologists for well over a hundred years. It has resisted attempts to classify it within the established groups of fishes. Over the years it has been suggested that it is an elasmobranch, or an agnathan, or a placoderm. Other suggestions have included a larval *Coccosteus*, a larval amphibian, and a larval lungfish. This last suggestion has received support from Thomson *et al.* (2003). However, Newman and Den Blaauwen (2008) disagree, and the debate continues. It still seems surprising that a larval form should have such a well-developed vertebral column, and show no changes in development with increase in size. There are no specimens known that shed light on how it might have metamorphosed into a small *Dipterus*; the only lungfish in the Achanarras fauna.

Curiously, this little fish is common at Achanarras, and very rare elsewhere. It is unknown outside Scotland. In most text-books *Palaeospondylus* merits a small section to itself, and maybe it should be given a new Class all to itself.

80 A well-preserved specimen of the enigmatic vertebrate *Palaeospondylus*.

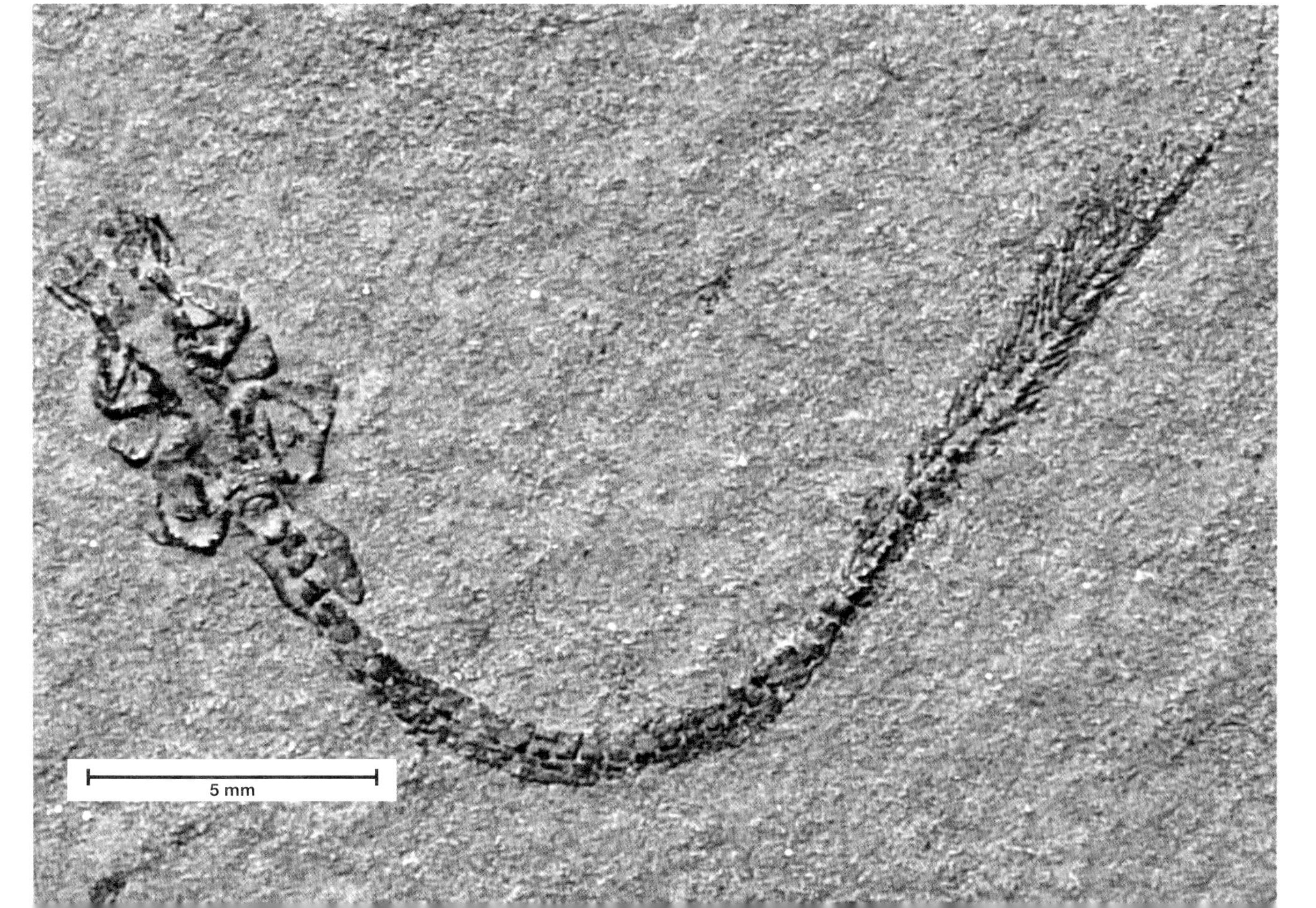

81

Dipterus / Fish / Devonian

Subphylum	**Vertebrata**
Class	**Osteichthyes**
Subclass	**Sarcopterygii**
Gen et sp	***Dipterus valenciennessi***
Locality	**Achanarras Quarry, Caithness**
Age	**Mid-Devonian**
Stratigraphy	**Caithness Flagstone Group, Middle Old Red Sandstone**

Dipterus is a lobe-finned fish, having pectoral and pelvic fins with a scaled central lobe with lateral fin rays. It is a lungfish, related to surviving lungfish still found in Australia, Africa and South America. Modern lungfish can gulp air to help them survive when the oxygen content of water is low, and they can also aestivate when water dries up. They dig a burrow and seal themselves inside, reducing body functions to near zero. They emerge at the time of the next flood. We do not know whether *Dipterus* could aestivate, but it does seem to be associated with conditions where other fish did not live, so maybe it could better withstand low oxygen conditions.

This is one of the commonest fish at Achanarras Quarry, and many are superbly preserved, allowing accurate reconstructions of the fish to be made (Ahlberg and Trewin 1995). On some bedding planes in the quarry the *Dipterus* specimens are all of very similar size; this may be due to a particular year-class shoal of the fish dying at the same time in mass-mortalities.

81a The lungfish *Dipterus* showing the pectoral lobe-fin flat against the body.

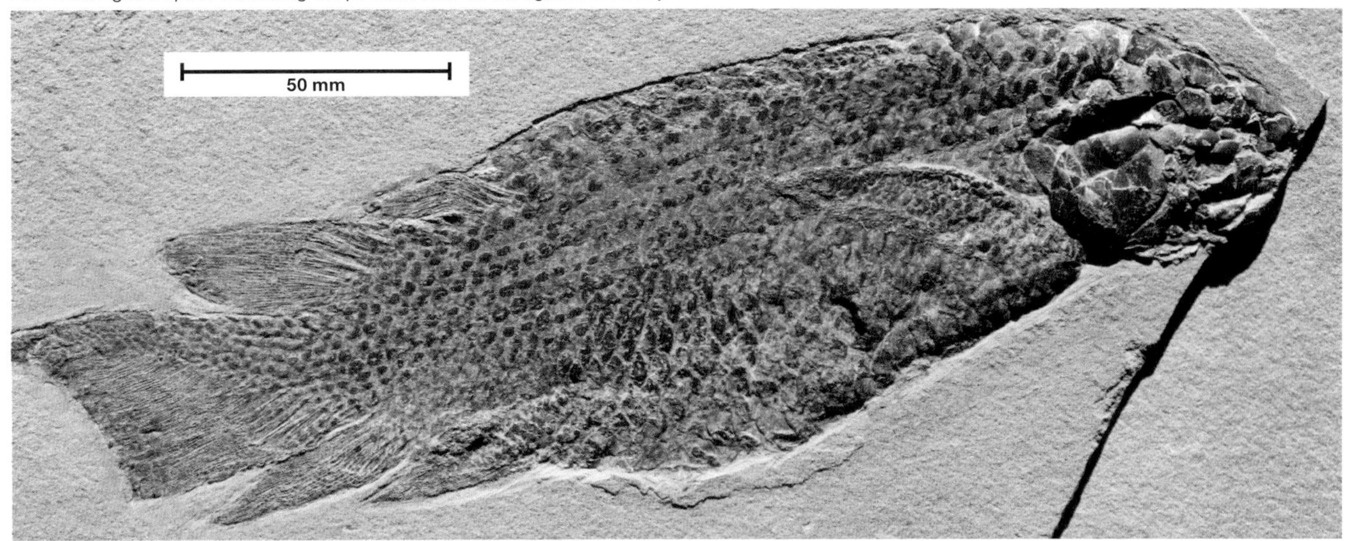

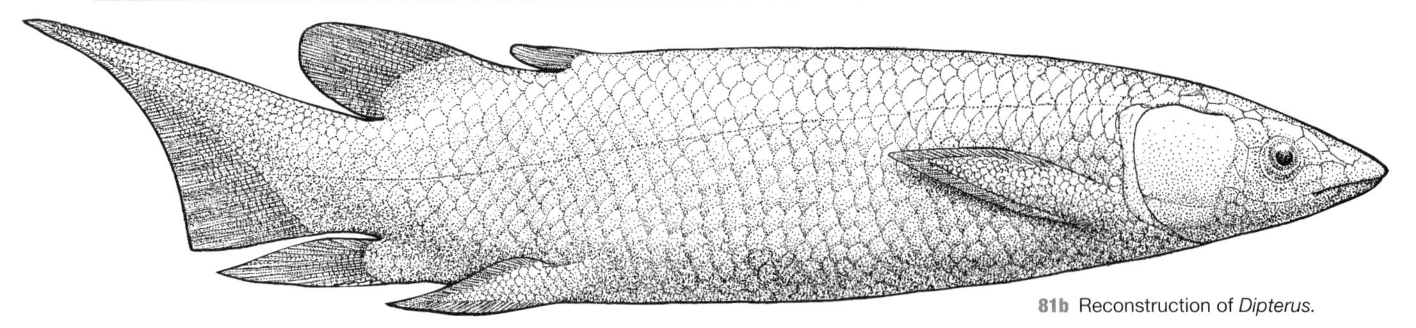

81b Reconstruction of *Dipterus*.

82

Osteolepis / Fish / Devonian

Subphylum	**Vertebrata**
Class	**Osteichthyes**
Subclass	**Sarcopterygii**
Gen et sp	***Osteolepis macrolepidotus***
Locality	**Cruaday Quarry, Sandwick, Orkney**
Age	**Mid-Devonian**
Stratigraphy	**Stromness Flagstone Group, Middle Old Red Sandstone**

Osteolepis is a common fish in the Sandwick fish bed of the Middle Old Red Sandstone in Orkney, and is also found at other localities of equivalent age, such as Tynet Burn (Fochabers) and Cromarty (Black Isle), but it is rare at Achanarras Quarry in Caithness.

This is a lobefin fish, related to lungfishes (see *Dipterus*) and coelacanths. *Osteolepis* has a typical heterocercal tail, two dorsal fins, an anal fin and paired pectoral and pelvic lobed fins. The scales are enamelled and rhomb-shaped. The bones of the skull are organised like those of later amphibians, the first land-living vertebrates, and it is from this group of fishes that amphibians, and from them, reptiles and mammals evolved.

There is a variety of osteolepid fishes known from the Orcadian Basin area, including *Thursius* and *Gyroptychius*. Jarvik (1948) published a classic monograph of this group of fish, and a short field guide was produced by Jack Saxon (1975) who studied and collected these fish from his home base in Thurso.

82a *Osteolepis* from the Sandwick fish bed, Orkney.

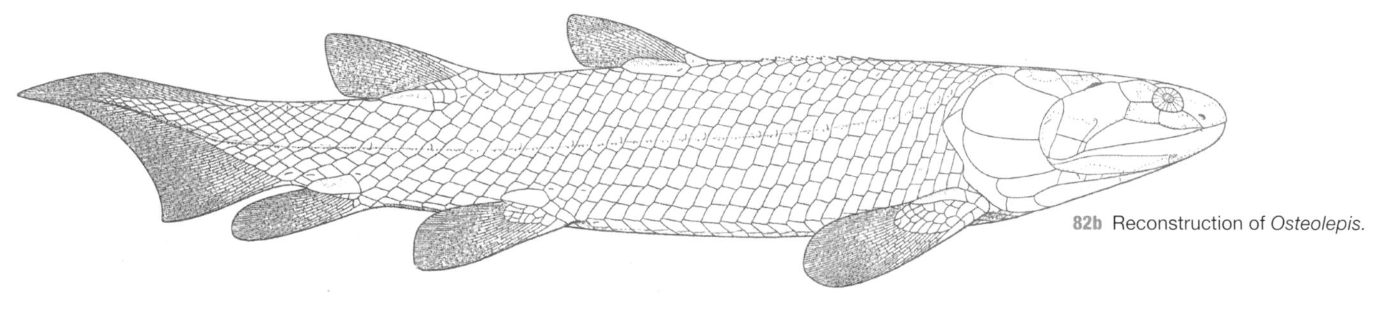

82b Reconstruction of *Osteolepis*.

83

Holoptychius / Fish / Devonian

Subphylum	**Vertebrata**
Class	**Osteichthyes**
Subclass	**Sarcopterygii**
Gen et sp	***Holoptychius***
Locality	**Dura Den, Fife**
Age	**Late Devonian**
Stratigraphy	**Dura Den Formation, Upper Old Red Sandstone**

The presence of well-preserved fossil fish at Dura Den was recorded in 1836 when a complete specimen of *Holoptychius* was found, and later described and figured by Agassiz in *Poissons Fossiles du Vieux Grès Rouge* as *Holoptychius andersoni*, named after the Revd John Anderson, who in 1859 published a monograph on the fossil fishes found at Dura Den. Anderson records visits to Dura Den by Lyell, Malcolmson and Hugh Miller, but it was in 1858 that spectacular finds were made at the time of a visit by a party including Murchison, Lord and Lady Kinnaird and a 'distinguished party from Rossie Priory'. A fine *Holoptychius* over 3ft long was found, and numerous smaller specimens. Later the owners of the quarry (who must have seen a profit!) cleared a large surface from which Anderson records that 'nearly a thousand fish were lifted from their stony bed of ages'. This event inspired Anderson to write his Monograph, clearly a fairly rushed project by today's standards.

The sandstones of Dura Den were deposited partly by rivers and partly as wind-blown sand dunes. In many specimens it can be demonstrated that fish carcasses dried out and were filled with wind-blown sand. Hence it is likely that the fish were trapped in pools that dried out, and the carcasses then covered by sand. Possibly floods had forced the fish out of river channels, but they became isolated in pools as the flood subsided.

Other concentrations of fish have been found at different levels in the sandstone, notably a surface with many *Bothriolepis*. Slabs of sandstone with Dura Den fish were clearly distributed far and wide at the time of discovery, as examples seem to be present in most of the older local museums and universities in Scotland, and magnificent specimens reside in the Natural History Museum in London.

83a *Holoptychius* crowded in sandstone from Upper Old Red Sandstone of Dura Den, Fife.

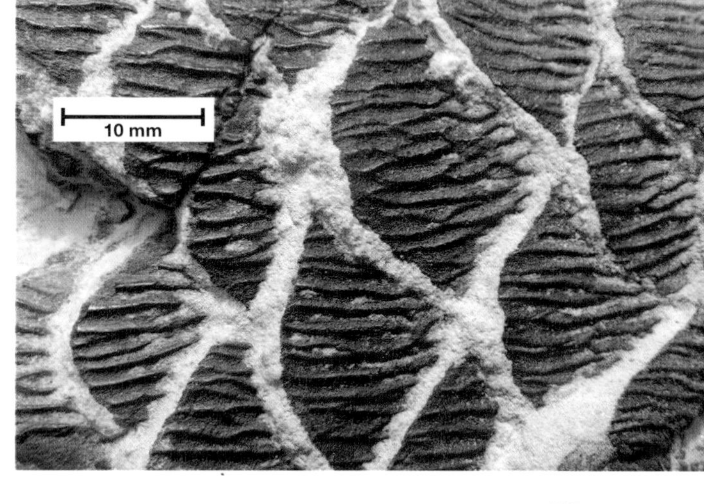

83b Typical ornament of *Holoptychius* scales.

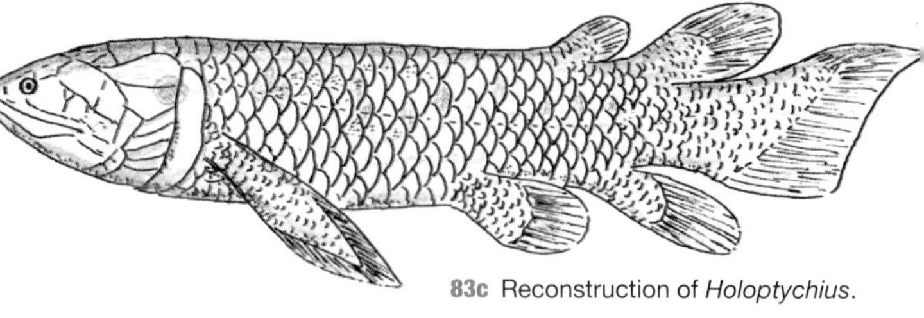

83c Reconstruction of *Holoptychius*.

84

Cheirolepis / Fish / Devonian

Subphylum	**Vertebrata**
Class	**Osteichthyes**
Subclass	**Actinopterygii**
Gen et sp	***Cheirolepis trailli***
Locality	**Achanarras Quarry, Caithness**
Age	**Mid-Devonian**
Stratigraphy	**Caithness Flagstone Group, Middle Old Red Sandstone**

Cheirolepis is a ray-finned fish, generally considered to an ancestor of the modern bony, (teleost) fish that include our common table fishes such as tuna, cod and trout. The fish has the primitive heterocercal (upturned) tail, and the body is covered with tiny scales. The jaws have small sharp teeth and the fish also had large eyes. The structure of the skull and jaws allowed it to open its mouth very wide, and it was probably a predator, able to swallow fish over half its own length. With large dorsal, anal and caudal fins and a streamlined body, *Cheirolepis* was a powerful swimmer. However its pectoral fins, used for steering, were not easily manoeuvrable, and it may not have been able to twist and turn easily in a chase.

This fish is one of the less common fish at Achanarras Quarry, but the specimens are usually very well preserved, and make *Cheirolepis* the best-known Devonian actinopterygian fish. This group of fish diversified rapidly in the late Devonian, and has a world-wide distribution.

84a *Cheirolepis* specimen with partly disarticulated head.

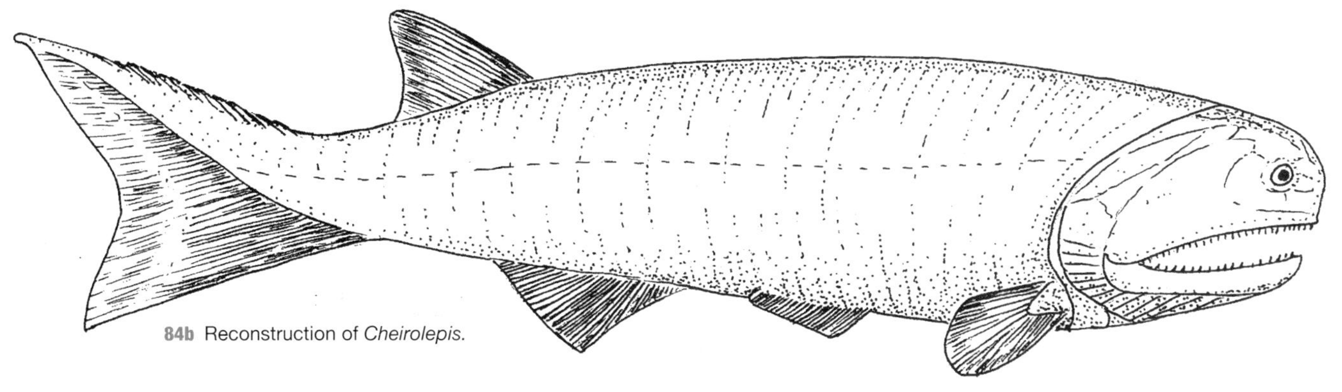

84b Reconstruction of *Cheirolepis*.

85

Rhizodus / Fish / Carboniferous

Subphylum	**Vertebrata**
Class	**Osteichthyes**
Subclass	**Sarcopterygii**
Gen et sp	***Rhizodus hibberti***
Locality	**Gilmerton, Edinburgh**
Age	**Carboniferous**
Stratigraphy	**Coal Measures**

Rhizodus was a fearsome beast that lived in the rivers of the Carboniferous coal swamps. Fossils of this animal are fragmentary, and it was usually found in coal mines where excavation of fish fossils was not a priority. Hence we do not know a lot about the detailed anatomy of this fish, and have to rely on fragmentary remains. The most impressive feature was the size; the lower jaw can be up to a metre long, and this gives a predicted length of the fish as 5–8 metres. It had curved, laterally compressed, dagger-like teeth, with several very large teeth up to 20cm long. With such weapons it must have been the top predator, and probably preyed on the larger amphibians of the coal swamps. The size of the animal suggests that the rivers in the coal swamps were wide with deep pools. One can imagine *Rhizodus* ambushing amphibians in the muddy waters.

Apart from impressive specimens of jaws with teeth, isolated teeth and the large shiny scales are sometimes found in coal seams. The illustrated specimen from the Dundee Museum collection is a typical jaw from a small individual. *Rhizodus* is quite closely related to *Osteolepis* of the Devonian period.

85 Jaw of the giant Carboniferous fish *Rhizodus*.

86

Akmonistion / Fish / Carboniferous

Subphylum	**Vertebrata**
Class	**Chondrichthes**
Gen et sp	***Akmonistion zangleri***
Locality	**Bearsden, Glasgow**
Age	**Carboniferous, Namurian**
Stratigraphy	**Manse Burn Formation**

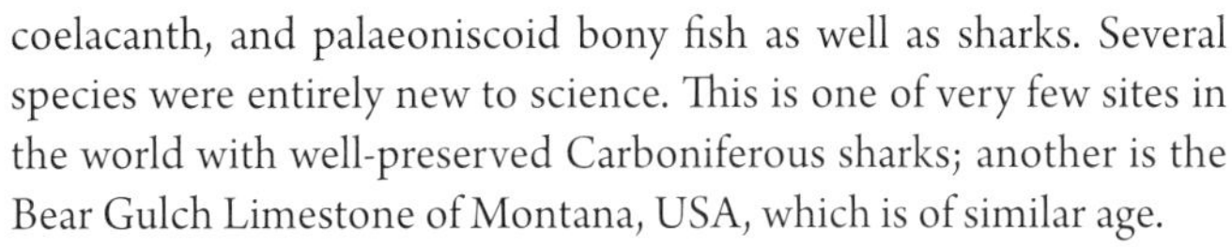

Akmonistion is a very unusual shark, bearing a strange spine that is modified into a dish-shaped structure covered with denticles that look like teeth. Similar denticles were also present on the top of its head, making it look as though it had a radar dish mounted on its head. The bizarre appearance of this fish has generated a lot of publicity, and numerous reconstructions and film animations exist on the Web, generally under the name *Stethacanthus*. It was described in detail by Coates and Sequeira (2001) who erected the new genus *Akmonistion* for this shark. This is one of many sharks and other fish found by Stan Wood in 1981, and excavated in his 'Bearsden Shark Dig' (Wood 1982). Stan had found sharks in ironstone nodules in the Wardie Shales near Edinburgh in 1971, and *Diplodoselache woodi* was named in his honour. Stan moved house to Glasgow and amazingly found fossil sharks close to his house in Bearsden, and organised a dig which revealed a rich fauna with 14 species of fish including acanthodians, a coelacanth, and palaeoniscoid bony fish as well as sharks. Several species were entirely new to science. This is one of very few sites in the world with well-preserved Carboniferous sharks; another is the Bear Gulch Limestone of Montana, USA, which is of similar age.

Sharks have skeletons of soft cartilage rather than bone, and thus complete specimens of sharks are rare – all that is usually found as fossils are teeth and fin spines. The organic-rich shales in which the Bearsden sharks are preserved were deposited in anoxic conditions where there were no scavengers, and the carcasses were buried in mud before they could decay. Both marine and freshwater fossils are found, so this may have been a lagoon environment with intermittent connection to the sea. There has been much speculation on the function of the strangely modified spine of this shark. Suggestions have been made that it was an electric organ to stun prey, a device to attach to a larger fish to take a tow as remora do today, a defensive organ, or that it was connected with mating behaviour. However, it is difficult to see how it could have been employed.

86a The 'Bearsden shark', *Akmonistion.*

86b Reconstruction of *Akmonistion.*

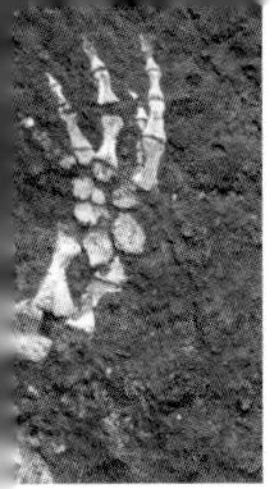

Vertebrates – Amphibians

Amphibians evolved from osteolepiform fish during the Devonian. The oldest tetrapod trackways are mid-Devonian, but in Scotland there are rare tetrapod trackways in the late Devonian Upper ORS of Tarbat Ness on the Black Isle (Rogers 1990). *Elginerpeton* is known from fragmentary remains found near Elgin and has tetrapod features. The late Devonian animals *Acanthostega* and *Ichthyostega* from Greenland retained a fish-like tail and had well-developed feet, but it seems they were suited more for swimming than walking, having weak wrist and ankle joints. They also retained internal gills, and it is clear that the animal was not fully terrestrial, but probably pushed its way through weedy shallows in search of prey.

True tetrapods suited to life on land appear in Scotland in the early Carboniferous. *Pederpes* is an important find, since it dates from the earliest Carboniferous, a period virtually devoid of tetrapod remains. Its main claim to fame is the earliest foot adapted to walk on land.

The most remarkable Carboniferous tetrapod fauna comes from East Kirkton in West Lothian. *Balanerpeton* is described below, but several others have been described, ranging from large animals that probably ate fish and other amphibians, to small lizard-like forms that were fully terrestrial. One, *Westlothiana*, was originally described as the oldest reptile or amniote in the world, but more recent evidence suggests it lacks the full features to be considered an amniote. It seems that the aquatic near-amphibians of the late Devonian became adapted to the wet swampy environments of the Carboniferous forests. Some remained aquatic, but others left the water and exploited as a food source the great variety of arthropods, including the early insects, that abounded in the forest.

87

Elginerpeton / Amphibian / Devonian

Subphylum	**Vertebrata**
Class	**Amphibia, Elginerpetontidae**
Gen et sp	***Elginerpeton pancheni***
Locality	**Scaat Craig, near Elgin**
Age	**Late Devonian**
Stratigraphy	**Upper Old Red Sandstone**

This is an isolated incomplete jaw from an animal that is one of several now known that have features intermediate between fish and amphibians. This jaw, together with a humerus and a tibia, have features that Ahlberg (1995) concluded indicate that the animal was more closely related to tetrapods than to fish. This specimen is from the late Devonian, and together with other finds of similar age from around the world, it has filled part of an information gap that existed concerning the evolutionary transition from fish to amphibian. In the past 30 years at least six new animals have been found that show features intermediate between fish and amphibians.

This jaw bone of *Elginerpeton* is worn by transport in a river system, and was found in coarse pebbly sandstone along with fish remains that fix the age of the deposit. This and other finds of early tetrapods shows that a lot of progress has been made in the past 20 years to fill gaps in the palaeontological record between fish and tetrapods.

87 The partial jaw of *Elginerpeton*.

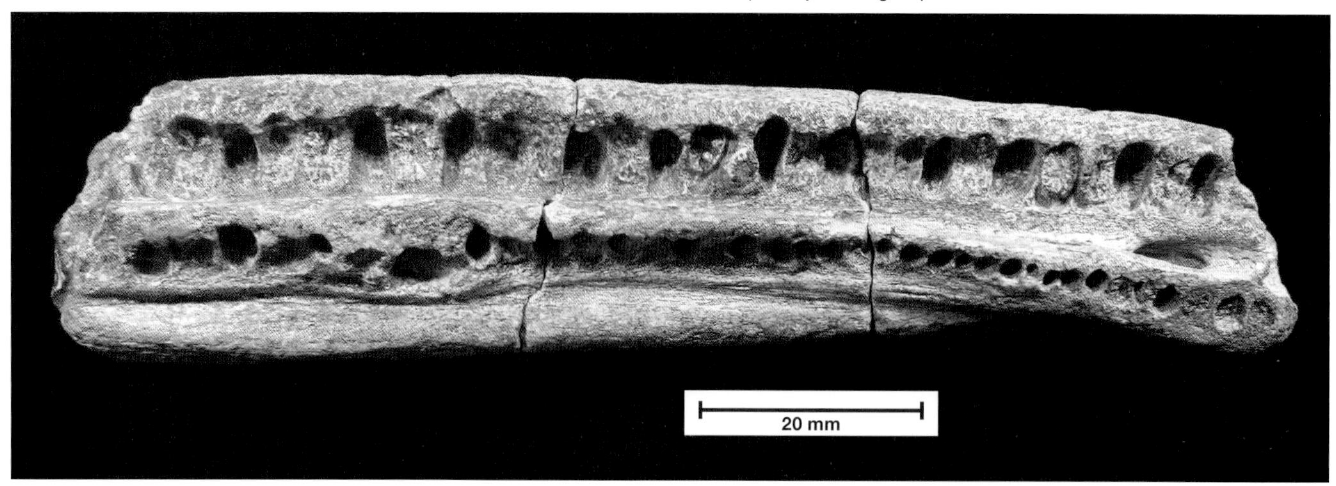

88

Balanerpeton / Amphibian / Carboniferous

Subphylum	**Vertebrata**
Class	**Amphibia**
Gen et sp	***Balanerpeton woodi***
Locality	**East Kirkton Quarry, Bathgate, West Lothian**
Age	**Carboniferous**
Stratigraphy	**East Kirkton Limestone**

Balanerpeton and *Westlothiana* (*see* below) are representatives of the tetrapod fauna of the Carboniferous of East Kirkton Quarry discovered by Stan Wood in 1984. This temnospondyl amphibian is the most common tetrapod at East Kirkton. The animal, as described in detail by Milner and Sequeira (1994) grew to a length of about 50cm, but most specimens are much smaller, and a possible larval form has been found in ostracod-rich shale. The adult had well ossified limbs and an ear structure indicative of a terrestrial habitat, but its abundance suggests it favoured the water margin as a habitat. However, it is worth remembering that modern newts travel a long way from water out of the breeding season, despite having a tail clearly adaped to an aquatic way of life.

The skull has large orbits that held eyes giving good lateral vision. Its jaws had numerous small sharp teeth, with several longer teeth suitable for piercing and holding prey; thus *Balanerpeton* appears to have been a carnivore. In overall appearance it resembles the modern salamander *Dicamptodon*, but with better lateral vision. The abundance of this amphibian suggested to Clarkson *et al.* (1994) that it probably lived around the shores of the East Kirkton lake, and laid its eggs in the shallows. Fish are absent or rare in the beds with *Balanerpeton*, thus the absence of fish predators may have made East Kirkton an attractive breeding pond and a safer place for the larval tadpoles of this amphibian.

88a The amphibian *Balanerpeton* from East Kirkton.

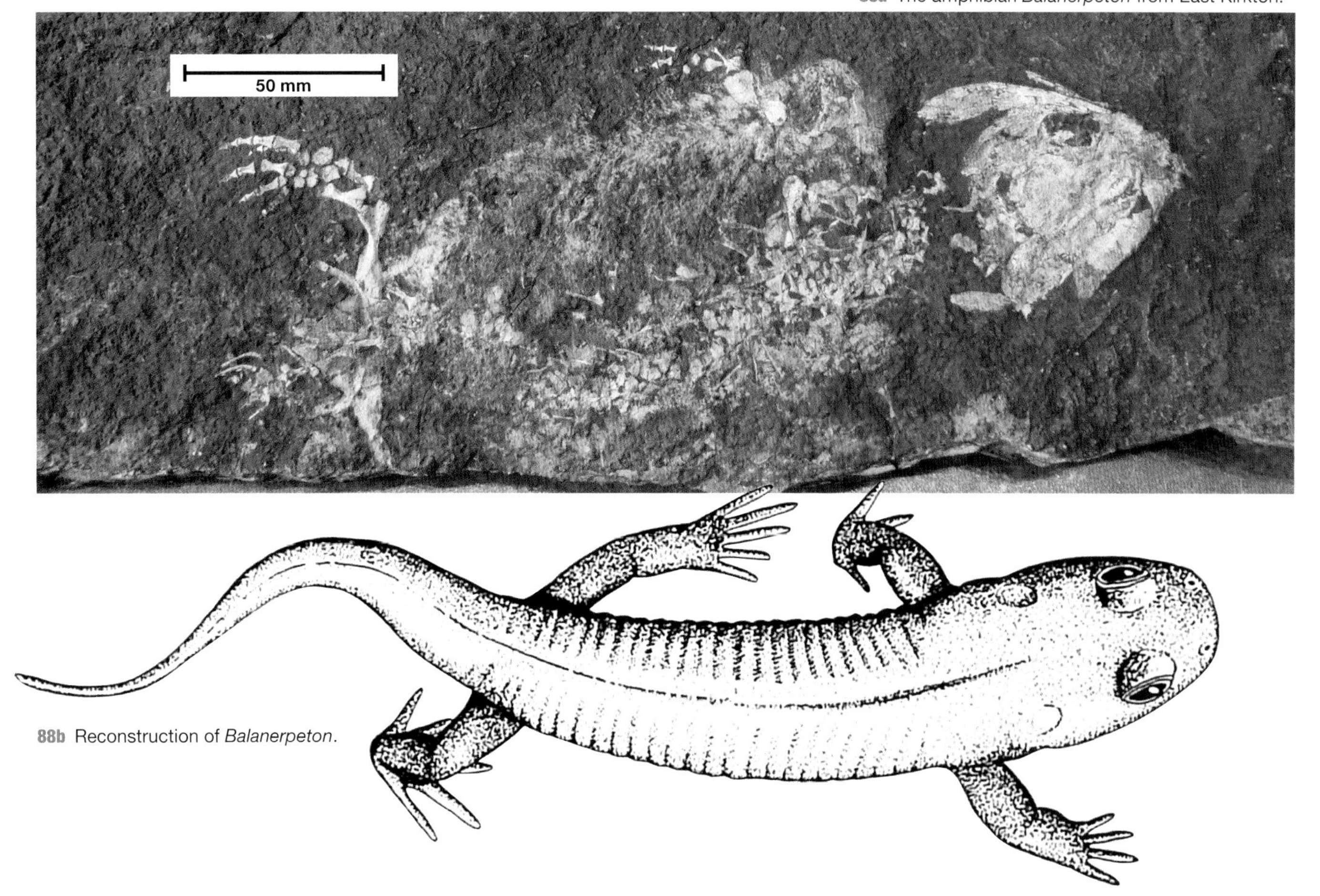

88b Reconstruction of *Balanerpeton*.

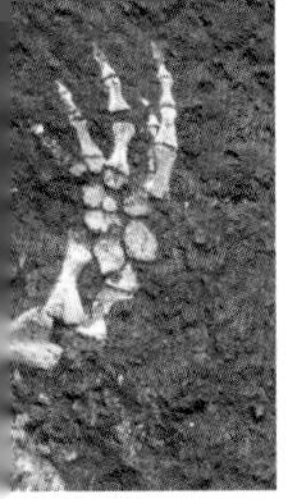

89

Pederpes / Amphibian / Carboniferous

Subphylum	**Vertebrata**
Class	**Amphibia**
Gen et sp	***Pederpes finneyae***
Locality	**Dumbarton, nr. Glasgow**
Age	**Carboniferous, Tournaisian**
Stratigraphy	**Ballagan Formation, Inverclyde Group**

Pederpes was the first tetrapod skeleton found from the earliest Carboniferous (Tournaisian) and shows the earliest known foot that is adapted for walking on land. There had been a gap in the geological record between the aquatic amphibians of the late Devonian, and the terrestrial animals of the mid-Carboniferous. This is known as 'Romer's Gap', and thus *Pederpes* is important in helping to bridge what was a gap in the tetrapod record at a time when they were evolving for a life on land (Clack 2002). The only known specimen was fully described by Clack and Finney (2005). Other recent finds in Scotland will add to knowledge of the earliest Carboniferous tetrapods, and aid in filling 'Romer's Gap'.

The specimen was 'discovered' in a store at the Hunterian Museum in Glasgow by Jon Jeffery, a student studying Carboniferous fish, including *Rhizodus*. He recognised the specimen as unusual and took it to his laboratory in Cambridge, where he and his supervisor, Jenny Clack, rapidly realised that this was not a rhizodont fish but a tetrapod. It had a label indicating it came from the 'Calciferous Sandstone Series', and by extracting spores from the rock matrix John Richardson of the Natural History Museum was able to assign a Late Tournaisian age, and hence *Pederpes* became the oldest tetrapod known from the Carboniferous. It transpired that the specimen had been found by Peder Aspen, after whom it is now named, and that it came from the Ballagan Formation

Thus a chance observation on a specimen that had been lurking in the museum for 26 years, and had been missed by several experts, produced a highly significant fossil. When discovered, the specimen was encased in a nodule with only parts showing; the careful preparation of the specimen took place over a period of four years.

89a The amphibian *Pederpes* after preparation.

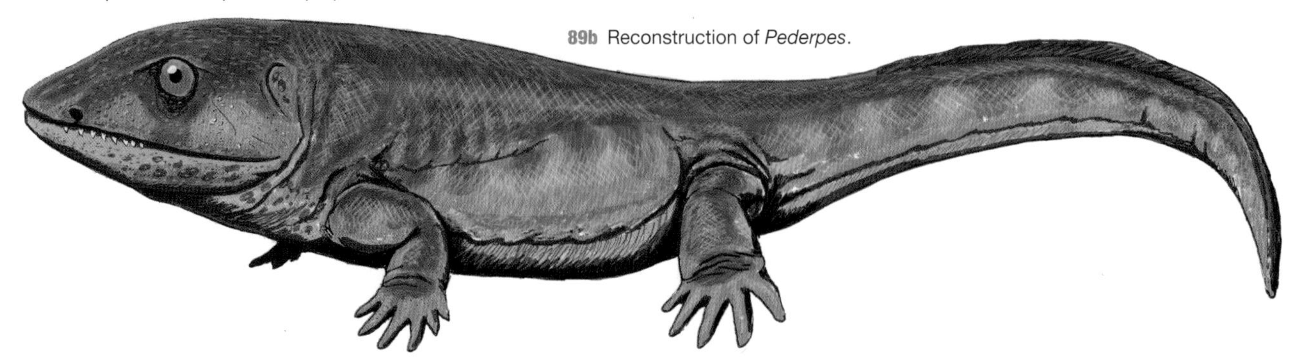

89b Reconstruction of *Pederpes*.

90

Westlothiana / Amphibian / Carboniferous

Subphylum	**Vertebrata**
Class	**Amphibia, Reptiliomorpha**
Gen et sp	***Westlothiana lizziae***
Locality	**East Kirkton, West Lothian**
Age	**Carboniferous, Brigantian**
Stratigraphy	**East Kirkton Limestone, Bathgate Hills Volcanic Formation**

In the early Carboniferous the famous East Kirkton site was situated in an area of active volcanicity. The East Kirkton Limestone is associated with volcanic tuffs and also limestones and cherty deposits that are probably the deposits of hot springs entering a lake through fissures. The lake margin was forested, and terrestrial animals and plants were preserved in the lake sediments, along with aquatic animals such as fish and eurypterids. The site is an old quarry, and was known in the 1830s, but it was the collector Stan Wood who started finding amphibian fossils at the site in 1984. His first finds came from a drystone dyke which he purchased from the landowner. Later he excavated in the old quarry over several years and collected an impressive range of fossil amphibians, fish, eurypterids, scorpions and a millipede.

Westlothiana – the name sponsored by West Lothian Council – was originally nick-named 'Lizzie the lizard' and hailed as the oldest reptile in the world (Smithson and Rolfe 1990), and it certainly has some reptilian features. However, after further preparation of the specimen, Smithson *et al.* (1994) concluded that it does not have all the features required to make it a reptile, and it is now classed as a reptiliomorph, a subdivision of the amphibians. As more fossils are found and studied it becomes increasingly difficult to draw a line between amphibians and reptiles. It is only to be expected that intermediate forms evolved that were transitional to reptiles – examples of the so-called 'missing links' in evolution.

Fossils of early Carboniferous tetrapods are very scarce, thus the East Kirkton site is very important in the study of tetrapod evolution. The unusual conditions favouring the preservation of small animals that fell or were washed into the hot spring lake. The reconstructed scene by Mike Coates (from Clarkson *et al.* 1994) shows Westlothiana looking out over the East Kirkton Lake in a landscape of dense forest with active volcanoes.

90a *Westlothiana*, 'Lizzie' the proto-reptile from East Kirkton.

90b Reconstruction of *Westlothiana* in a scene from the Carboniferous of East Kirkton.

Vertebrates - Reptiles

Amniotes differ from the amphibians in that they lay eggs with a shell that contains enough food for the embryo to emerge as an air-breathing terrestrial animal – thus avoiding the need to lay eggs in water, and not having an aquatic larval stage (tadpole). In the late Carboniferous and Permian the climate became dry, and an ability to survive out of water favoured those tetrapods that could sustain a fully terrestrial lifestyle; thus the reptiles evolved to exploit the new environments.

The most famous reptile faunas in Scotland are those that occur in Permian and Triassic sandstones in the Elgin area. At the time they were found all the sandstones in the Elgin Area had been mapped as Old Red Sandstone of Devonian age; thus the discovery of reptiles caused much excitement and discussion, and exchanges were not always harmonious when reputations of famous men were at stake. It was the little reptile *Leptopleuron* that caused the most academic trouble before its true Triassic age was established. We now know that there are two reptile faunas present in the Elgin area, one of late Permian age and the other early Triassic.

In the Jurassic, dinosaurs roamed in Scotland, and fragmentary fossils of several types have been found in Skye since 1995. Ichthyosaurs, plesiosaurs and turtles swam in the seas, but fossils of these marine reptiles are rare in Scotland when compared with the abundance of material found in the Jurassic of England. This difference is probably due to the fact that the area of outcrop of Jurassic strata in Scotland is small, and that Scottish Jurassic cliffs are not being as rapidly eroded as those of Dorset and Yorkshire in England. Furthermore, there has been a great deal of quarrying activity in the English Jurassic, particularly shales for brick manufacture, and limestones for building stone and lime production, so a lot more rock has been examined, resulting in numerous finds.

91

Gordonia / Reptile / Permian

Subphylum	**Vertebrata**
Class	**Reptilia**
Gen et sp	***Gordonia traquairi***
Locality	**Clashach Quarry, Hopeman, Moray**
Age	**Late Permian**
Stratigraphy	**Hopeman Sandstone Formation**

Clashach Quarry at Hopeman is well known for the variety of reptile trackways that have been found during quarry operations. A display of sandstone slabs with trackways can be seen near the entrance to the quarry. Most trackways are found in sandstones that were deposited as wind-blown sand dunes (*see* trace fossil *Chelichnus*, 103) However, no skeletal remains of reptiles had been found there until 1997, when a block of stone to be loaded on a lorry was split. A hole was seen in the stone – a flaw that caused the block to be rejected. Fortuitously, Carol Hopkins had shown the quarrymen specimens at Elgin Museum, where bones are only seen as hollow moulds in the sandstone. They saved the stone and Carol used a wire probe to discover that the hole had a regular shape, and probably represented dissolved bone.

Neil Clark of the Hunterian Museum was contacted, and the stone containing the hole was then cut to a reasonable size, and studied using x-ray techniques and a medical scanner. A computer-generated 3D image of the hole was then produced, revealing in extraordinary detail the skull and lower jaw of a dicynodont reptile. It has been referred to the genus *Gordonia,* and also to *Dicynodon.* The hole was all that was left of a reptile skull that had been completely dissolved by water percolating through the rock.

50 mm

91a The Elgin reptile *Gordonia*; a 3D replica skull.

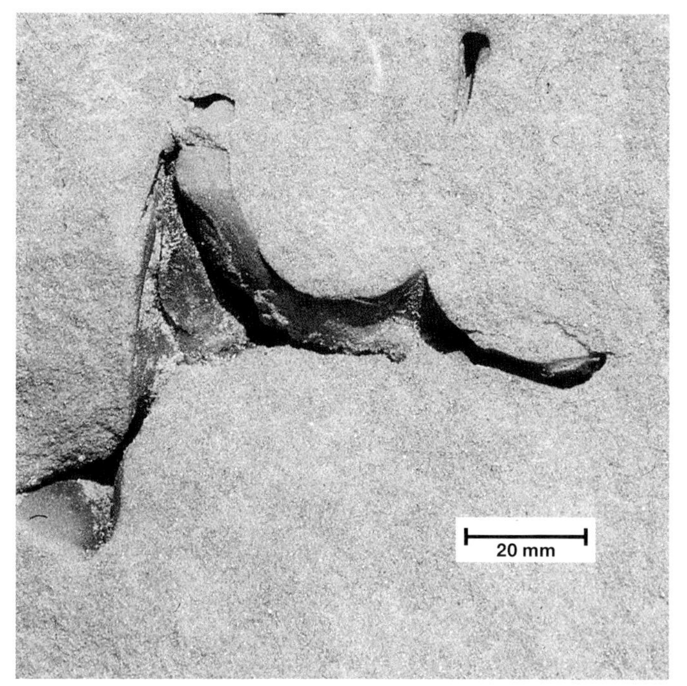

91b The original find; a hole in the rock that revealed the *Gordonia* skull.

The specimen illustrated is a resin model made from the MRI scan data (Clark 2008), together with the original find of the hole in the rock. Before the advent of this method of study a mould would have been made by pouring plaster or a rubber solution into the hole and allowing it to set; then the rock generally had to be destroyed to reveal the cast, which may not have been perfect. The new method avoids destruction of the original specimen, and reveals much greater detail. Doubtless numerous specimens have been destroyed by quarrying in the past because the significance of a hole in the Hopeman Sandstone was not recognised.

Gordonia had strong tusks that it probably used for digging, and it probably had a horny beak for cropping plants. It was the size of a small pig with a short tail, and is similar to late Permian reptiles that were common worldwide; thus the Hopeman Sandstone can be confidently dated to the late Permian.

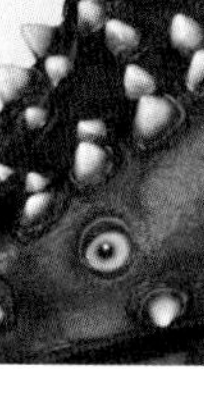

92

Elginia / Reptile / Permian

Subphylum	**Vertebrata**
Class	**Reptilia**
Gen et sp	***Elginia mirabilis***
Locality	**Cutties Hillock, Elgin, Moray**
Age	**Late Permian**
Stratigraphy	**Hopeman Sandstone Formation**

Elginia is only known from fragments from Cutties Hillock, but enough is known to reconstruct the animal. The skull is unusual in that it bears 16 spines, the largest looking like cow's horns. The skull is only 13cm long and the whole animal less than a metre long. It has short, strong legs and only a short tail. *Elginia* was a herbivore, but there are no plant fossils known from the Elgin sandstones. The reptiles were preserved under dune sands in areas lacking vegetation, but vegetation must have existed nearby, and would have comprised conifers, cycads and ferns. Doubtless there were valleys and pools with enough water to support vegetation. It seems the animals had to cross the dunes regularly, maybe on migration, or to visit a water hole. They left tracks in the dunes, and occasionally an animal died in the dunes and its remains were covered by blown sand and eventually fossilised.

92a *Elginia*, model of the horned skull (13 cm long).

92b Reconstruction of *Elginia*.

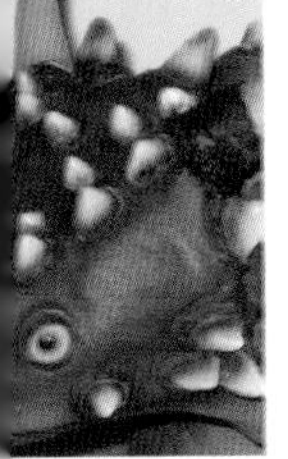

93

Stagonolepis / Reptile / Trias

Subphylum	**Vertebrata**
Class	**Reptilia**
Gen et sp	***Stagonolepis robertsoni***
Locality	**Lossiemouth**
Age	**Late Triassic**
Stratigraphy	**Lossiemouth Sandstone Formation**

This armoured reptile looks a bit like a crocodile with a very small head. Its back was covered in rectangular bony scutes with a distinctive ornament. The specimen illustated here was found in 1844 by a quarryman at Lossiemouth, who took it to Patrick Duff, town clerk of Elgin and a keen local geologist, who wrote a geological account of Morayshire (Duff 1842). The specimen shows a pattern of 31 scutes, but none of the internal bones. There is an early calotype portrait of Duff holding the specimen, and the specimen was also photographed using this early photographic process (*see* Andrews (1882) for details). Alexander Robertson, another local geologist, sent an image to Louis Agassiz who identified the scutes as scales of a large fish related to *Glyptolepis.* He named this 'fish' *Stagonolepis robertsoni* after Robertson, maybe assuming he was the finder. A description and illustration was published by Agassiz (1844) in his famous

93a *Stagonolepis*, the type specimen, interpreted as fish scales by Agassiz.

monograph on the fishes of the Old Red Sandstone. Agassiz copied the photograph by a lithography process, and Andrews (1982) suggested that this is the earliest published photograph of a fossil vertebrate.

Remembering that at that time the rocks in which it was found were presumed to be Old Red Sandstone, and given the appearance of the specimen as rhomboid scales, it is not surprising that Agassiz did not think beyond fish in his identification. The specimen is from a small *Stagonolepis*, hence the size of the 'scales' was reasonable for a large fish, and larger scales of *Holoptychius* (83) were already known. However, when the reptile *Leptopleuron* (94) was found in the same rocks, intense scientific debate ensued, which is summarised below.

Later, in 1857–8, associated scutes and bones of *Stagonolepis* were found, and it was described as a reptile by Huxley (1859). Other reptiles were also found that compared with those already known by Huxley from Triassic rocks; thus the fossils were demonstrating that not all the sandstones of the Elgin area belonged to the Old Red Sandstone.

Stagonolepis grew to a length of about three metres, with short legs and a long tail. The head was small for the size of the body and the teeth are simple pegs. The skull had a short upturned snout that might have been used for digging up roots, and it was probably mainly a vegetarian. The armour plating of scutes may have been defensive, since the remains of a bipedal carnivore called *Ornithosuchus* are also found in the Triassic rocks of the Elgin area. This animal was similar in size to *Stagonolepis*, and was probably the top predator in the area. This specimen is on display in Elgin Museum along with other *Stagonolepis* specimens.

93b Reconstruction of *Stagonolepis*.

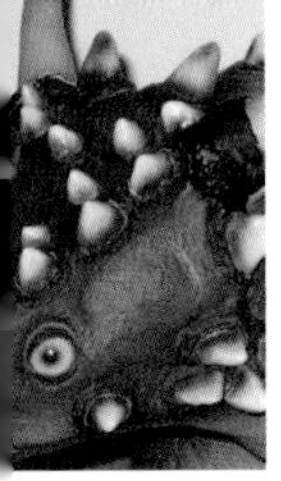

94

Leptopleuron / Reptile / Trias

Subphylum	**Vertebrata**
Class	**Reptilia**
Gen et sp	***Leptopleuron lacertinum*** (= ***Telerpeton elginense***)
Locality	**Spynie Quarries, Elgin, Moray**
Age	**Late Triassic**
Stratigraphy	**Lossiemouth Sandstone Formation**

This little Triassic reptile was found in 1851 and became the cause of much debate by the leading palaeontologists of the day. At the time it was found, the rocks in Spynie Quarry were thought to belong to the Old Red Sandstone, and if this had been true this little animal would have been the oldest reptile known. Also, such a find seemed to be contrary to the 'progressionist' views of many palaeontologists in the 1850s, that life had generally proceeded from simple to advanced forms through geological time, possibly through a series of creations (*see* Benton 1983). Progressionist ideas were to lead to ideas on evolution, culminating in the work of Darwin. Charles Lyell opposed the 'progression' of life and so initially welcomed this small reptile as support for his views.

This fossil also caused trouble in the geopolitics of the time. The fossil was obtained from a quarryman by Patrick Duff, geologist and Elgin Town Clerk. Captain Brickenden was Duff's brother-in-law, and he had collected specimens of dinosaur bones for Gideon Mantell, discoverer of *Iguanodon*, in southern England. Brickenden had found footprints of probable reptile origin in the Elgin area and had told Mantell; thus the small reptile was further evidence, and he wanted Mantell to describe it.

94a *Leptopleuron*, reptile skeleton in sandstone, preserved as an impression, the bone having dissolved.

The specimen was sent to London and Charles Lyell saw it on 1st December 1851, and had drawings made to go in a new edition of his *Manual of Elementary Geology*. He asked Mantell to describe the specimen, and by 5th December confirmed that Mantell would describe the specimen at a Geological Society meeting, and stated that Mantell would call it *Telerpeton elginense*. However, Richard Owen also saw the specimen, and it transpired that in November Duff had sent Owen the notice of the find that appeared in the *Elgin Courant*, before Mantell had heard of the find.

Mantell and Brickenden were due to put their paper before the Geological Society on 17th December, but the lecture was delayed. On 20th December Owen (1851) published an unillustrated description (dated 15th December) in the *Literary Gazette* in which he named the animal *Leptopleuron lacertinum*. Mantell (1852) described the reptile under the name *Telerpeton elginense*, but since Owen had published first, his name for the animal has priority.

Thus there was an unseemly rush to publish, and a feud that had developed over years between Owen and Mantell was intensified. It does look as though Owen pounced when the Geological Society meeting was delayed, or even anticipated that date by putting a date two days earlier on his description, and so giving his name priority. Such were the rivalries generated by this little reptile.

By 1859, more reptiles had been found in the Elgin area, and they were shown to be similar to Triassic forms known elsewhere. Lyell visited the area and later abandoned support for a reptile from the Old Red Sandstone. However, some local geologists did not accept the Triassic age, unwilling to give up the fame of having the oldest reptile fauna in the world.

What of the animal? It is a small, primitive reptile similar to a lizard; it has front teeth that are like chisels, which suggests that it was a herbivore, maybe having a lifestyle similar to modern rodents. It may also have preyed on small arthropods. It has recently been described in detail by Säilä (2010). Using the numerous specimens that have been found since 1851, she produced the excellent reconstruction of this small lizard-lke reptile reproduced here.

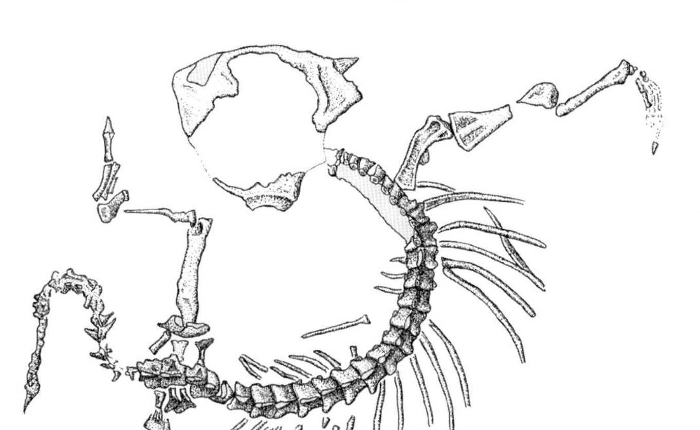

94b Drawings of the *Leptopleuron* specimen.

94c Reconstruction drawing of *Leptopleuron*.

95

Dinosaur, Sauropod / Reptile / Jurassic

Subphylum	**Vertebrata**
Class	**Reptilia, Dinosauria**
Gen et sp	**Possibly a Cetiosaurid type of Sauropod**
Locality	**Valtos, Isle of Skye**
Age	**Middle Jurassic**
Stratigraphy	**Valtos Sandstone Fm., Great Estuarine Group**

The first dinosaur fossil bones found in Scotland were described in 1995 in two papers in the Scottish Journal of Geology. One, from the Lower Jurassic Broadford Beds (Benton *et al.* 1995), is a tibia (leg bone) of a theropod dinosaur that was similar to *Coelophysis*. The other, shown here, was found in the Middle Jurassic near Staffin, and came from a sauropod dinosaur, possible a species of *Cetiosaurus* or close relative (Clark *et al.* 1995). More recently, Liston (2004) compared the bone with cetiosaur material from England and concluded that whilst the bone is a sauropod humerus, there is not enough evidence to assign it to a cetiosaurid. This dinosaur, about 11–13 m in length and with a long neck and tail, would have fed on vegetation in a lush, low-lying area with rivers, swamps and lakes.

95a Reconstructed limb bone of a sauropod dinosaur from Skye. The bone pieces are dark, and light colour highlights parts that are reconstruction.

20 cm

The specimen has an interesting history, having been reassembled from three parts. The first part was found by a party of geologists from BP on a field excursion, but it was clear that it had been damaged and part removed by a collector. Around the same time the other end of the bone was found very close to the site of the first find. Following publicity the 'missing piece' was posted anonymously to the Hunterian Museum, and the three parts were briefly reunited. The specimen is now in the Staffin Museum on Skye, along with other dinosaur finds from the area such as footprints (*see* trace fossils section). It is worth noting that the collector thought he had found a piece of fossil wood rather than a bone. It is not always easy to tell the difference in the field, so if in doubt don't damage the specimen, but report the find. It is rather disappointing to read (Liston 2004) that the proximal part of the bone has been retained by the original collectors. Sense should prevail; re-unite the original, and make casts for the three finders.

Further dinosaur finds have been made from Skye, all of isolated bones. One of particular interest is a specimen of part of the ulna and radius (arm bones) from a eurypodan dinosaur, the group that includes stegosaurs and ankylosaurs, in the Middle Jurassic Bearreraig Sandstone. This find, and the earlier find from the Lower Jurassic, occur in sediments of marine origin, and the bones must have come from carcasses washed into the sea from rivers and beaches.

95b Imaginary scene on Skye from the Jurassic; animals are generalised, based on the fragmentary specimens and trace fossils that have been found.

Vertebrates - Mammals

The earliest mammal fossils from Scotland are fragmentary remains from the Jurassic of the Isle of Skye, and are important fossils as representatives of some of the earliest mammals known, living alongside the dinosaurs. There is a general lack of mammalian fossils in Scotland, since we lack onshore sedimentary rocks of Tertiary (Palaeogene and Neogene) age. It is not until Quaternary times that significant mammal fossils occur in Scotland, found in beds of sand and peat between layers of till deposited by the great ice sheets that covered Scotland. These include the mammoth, giant elk, reindeer and woolly rhinoceros that were hunted by early man.

96

Mammoth / Mammal / Quaternary

Subphylum	**Vertebrata**
Class	**Mammalia**
Gen et sp	***Elephas primigenius***
Locality	**Woodhill Quarry, Greenhill, Kilmaurs, Kilmarnock**
Age	**Quaternary**
Stratigraphy	**Clay beneath till**

Mammoth remains, mainly in the form of isolated teeth and tusks, have been found at localities in southern Scotland ranging from Ayrshire and the Glasgow district in the west, to Eyemouth in the east. They are generally found when digging gravel, excavating foundations and making roads, and they occur in the unconsolidated sands, muds and peat deposited in association with till or boulder clay deposited from ice during the last ice age. The animals lived in tundra regions near the margins of the ice sheets. The teeth and tusks are the parts of the mammoth that are most resistant to decay and breakage; the spongy bone would have broken up and decayed quite rapidly.

The age of deposits containing mammoth bones has been radiocarbon dated at a number of sites, and the results give ages of around 25–30,000 years. Reindeer and woolly rhinoceros bones have also been found in these deposits.

96a Portion of the tusk of a mammoth.

96b Sketch reconstruction of mammoth.

97

Lynx / Mammal / Quaternary

Subphylum	**Vertebrata**
Class	**Mammalia**
Gen et sp	***Lynx lynx***
Locality	**Inchnadamph caves**
Age	**Quaternary**
Stratigraphy	**Cave deposits**

At Inchnadamph in the NW Highlands there is an extensive system of caves that formed in the Durness Limestone during warmer periods between ice ages. Animals used the caves as refuges and their bones are found scattered in the deposits of several caves. There is quite an age range represented, covering more than one interglacial period. The earliest are about 40,000 years old, and the youngest only 9000 years old, dating from the time of retreat of the last ice sheet. The only evidence for polar bear in Britain is from the caves; the specimen originally being identified as brown bear. These rather recent fossils show that Scotland had a fauna including brown bears, wolves, lynx, arctic fox and arctic lemmings. All these are now extinct in Britain, the wolf being the last to go through persecution by Man. There is a plaque in a layby between Brora and Helmsdale that records (allegedly) where the last wolf in Scotland was shot.

This is clearly an arctic fauna, and was typical of Scotland after the ice melted and forests became established around 10,000 years ago. Sea levels rose and the fauna was trapped on the British Isles, unable to migrate north with the climatic change. The lynx whose skull was found in the cave deposits would have hunted lemmings and probably mountain hares, and also taken deer calves and caught fish. It was probably hunted to extinction by Man, and human remains in the caves have recently been dated to about 4500 years BP. Thus the deposits in the caves link geology and archaeology in northern Britain following the end of the last glaciation

97 Skull of a lynx from Inchnadamph caves. Skull 15.5 cm long.

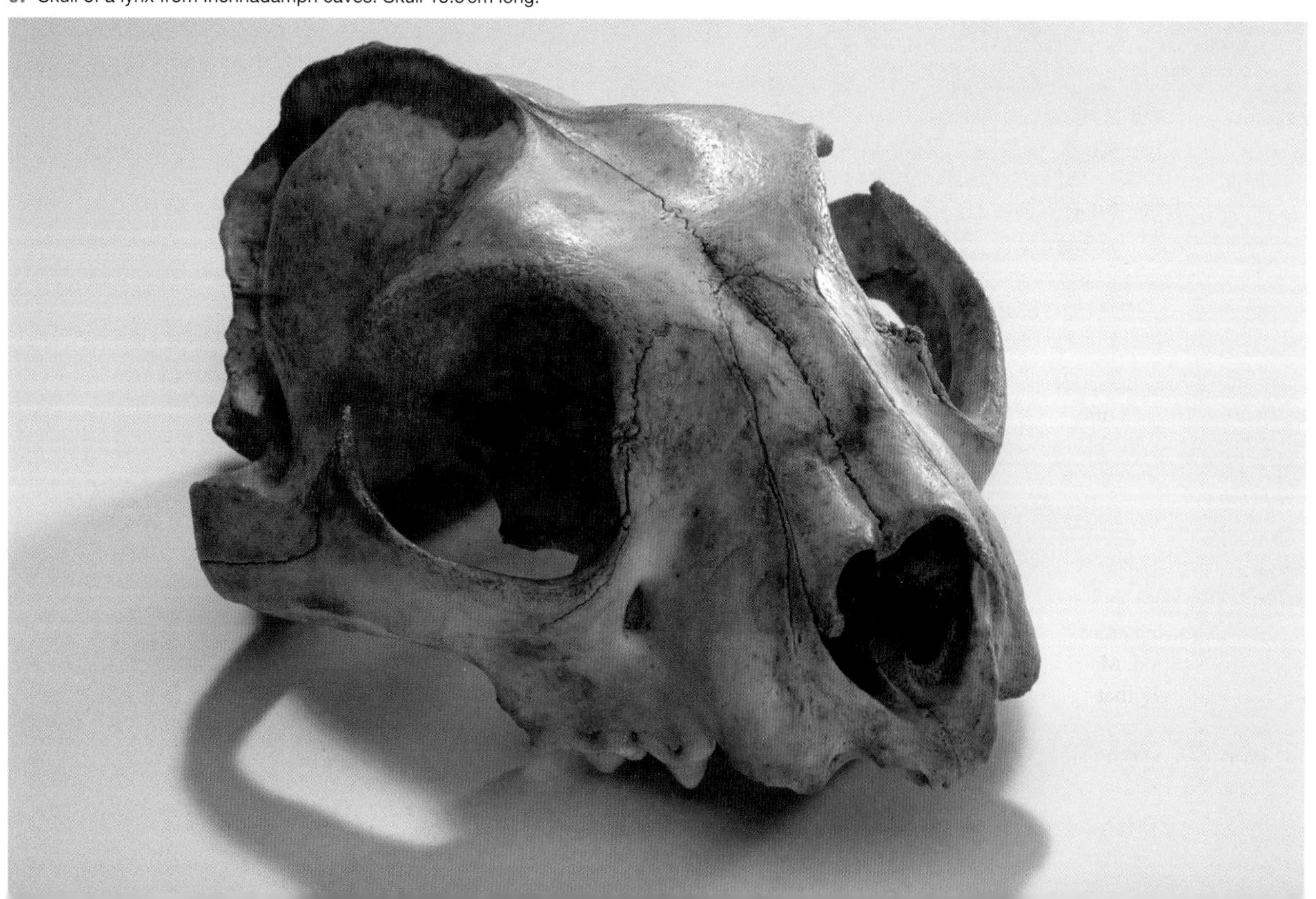

TRACE FOSSILS

Trace fossils are defined as 'evidence of the activity of an organism preserved in rock'. Thus dinosaur footprints, worm burrows, trilobite trackways, holes bored by bivalves, and bite marks left by predators are all examples of trace fossils produced by an activity of an animal, but do not contain any part (e.g. bone, shell) of the animal. Fossil faeces, known as coprolites, are also a category of trace fossil.

The study of trace fossils is known as ichnology, and much evidence can be gained both about the activities of animals such as worms that have not been fossilised, and about the environment in which they lived. Trace fossils do not belong to a phylum, and no natural classification exists. They have been grouped in several ways in the past. Groupings of traces characteristic of particular environments have been termed 'ichnofacies', and groupings based on the 'ethology', or type of activity (e.g. running, feeding, hunting, swimming, home building) have been attempted, but rely on our (frequently dubious) interpretations of traces. Thus trace fossils are frequently arranged in alphabetical order, and although they are given generic and specific names, these are not biological species, merely recognisable physical forms.

General assumptions are that any one animal can make a variety of trace fossils, and similar trace fossils can be made by different animals. The following section includes a few examples of trace fossils.

98

Skolithos (Pipe Rock) / Trace Fossil / Cambrian

Trace fossil

Gen et sp	***Skolithos linearis***
Locality	**Inchnadamph**
Age	**Early Cambrian**
Stratigraphy	**Eriboll Formation, Pipe Rock Member**

The trace fossil *Skolithos* is a vertical burrow. There are many types of *Skolithos*, varying in the diameter and depth of the burrow. In the NW Highlands the Cambrian succession includes a rock unit that is known as the 'Pipe Rock' on account of the general appearance given by the numerous vertical *Skolithos* burrows. The burrows stand out due to variable pink and red colour staining in the quartzite rock. On bedding surfaces the burrows are visible due to differences of hardness between the burrows and matrix, resulting in different weathering rates. The hard quartzite was originally sand in a shallow Cambrian sea, and the animals burrowed to produce a protective home, sheltered from predators and strong currents.

The burrowing animal is not preserved, and could have been a type of worm, or even an arthropod. Many different types of modern animals make vertical burrows, and it is unlikely that we will ever be sure what made these Cambrian *Skolithos*. Similar *Skolithos* 'pipe rock' is found worldwide in shallow marine rocks of Cambrian to Silurian age. Presumably the animals responsible for these deep vertical burrows died out around the end of Silurian time.

98 'Pipe rock' with *Skolithos* burrows, showing cut vertical section and bedding surface appearance.

99

Dictyodora / Trace Fossil / Silurian

Trace fossil	
Gen et sp	***Dictyodora scotica***
Locality	**Thornylee Quarry, near Innerleithen, Peeblesshire**
Age	**Early Silurian**
Stratigraphy	**Gala Group**

This is a burrow found in shales that were deposited in deep marine conditions in the early Silurian. When the strange meandering markings on the rock surfaces were first mentioned in 1850 they were initially thought to be impressions of fossil worms. However, it was soon realised that the markings were due to sediment disturbance by burrowing animals. The burrow system has been described by Benton and Trewin (1980) and consisted of a basal burrow from which a sharp ridge extends up into the sediment. The animal must have burrowed horizontally in the mud whilst keeping contact with the water above with a narrow organ that maybe supplied it with oxygen.

The tight, meandering form indicates efficient use of the mud for feeding; it only rarely crosses its own path and the arms of the meanders are close together. This is a form of feeding that developed in the Silurian, and has been adopted by many burrowing and grazing animals living in deep seas today.

99 *Dictyodora*, a tightly meandering trace made by an animal feeding in deep-sea mud.

100

Beaconites / Trace Fossil / Devonian

Trace fossil

Gen et sp	***Beaconites***
Locality	**New Aberdour, Aberdeenshire**
Age	**Early Devonian**
Stratigraphy	**Crovie Group, Lower Old Red Sandstone**

Beaconites is the name given to a large burrow that is widespread in Devonian sandstones that originated as sand deposited by river systems. The burrows are large, up to 13cm in diameter, and have a fill of curved packets of sand that were packed behind the animal as it burrowed through the sediment. At New Aberdour these burrows can be found in sandstones which represent fossil soils formed in a semi-arid climate (caliche soils), and also in pebbly sand with cross-bedding structures indicating deposition in small river channels. It is probable that the animal was a large arthropod looking rather like a large, 30cm long woodlouse. The animal probably scavenged vegetation on the surface, but dug burrows to escape from the heat of the sun, and to conserve moisture. The trace fossil trackway called *Diplichnites* (*see* 101 below) may represent the walking trace of the animal that made *Beaconites*. The field geology of the Old Red Sandstone at New Aberdour has been described by Trewin and Kneller (1987b).

100 *Beaconites*, a large burrow commonly found in fluvial deposits of the Old Red Sandstone.

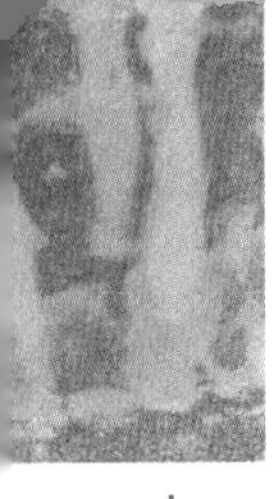

101

Diplichnites / Trace Fossil / Devonian

Trace fossil	
Gen et sp	***Diplichnites***
Locality	**Gamrie, Banffshire**
Age	**Early Devonian**
Stratigraphy	**Crovie Group, Lower Old Red Sandstone**

On the shore immediately west of the harbour wall at Gamrie there are a few sandstone beds that have trackways on the surfaces. The surface illustrated was photographed many years ago and may not still be visible. The tracks are 5 to 14cm wide and consist of numerous closely-spaced prints. Some specimens show curved sections of the trackway where the animal was turning, and it is possible to count the imprints made by the feet of the animal. There are at least 10 pairs of legs involved in walking, and all were similar. This indicates that the animal was a myriapod of some kind, distantly related to modern centipedes and millipedes. The animals lived at the margins of an alluvial plain that was periodically converted to a shallow lake during floods from rivers draining higher land to the south (Trewin and Kneller 1987a).

101 *Diplichnites* trails on a sandstone bedding surface. Made by an arthropod with many legs.

102

Siskemia / Trace Fossil / Devonian

Trace fossil

Gen et sp	***Siskemia***
Locality	**Carrick, Ayrshire**
Age	**Early Devonian**
Stratigraphy	**Lower Old Red Sandstone**

Siskemia is a trace fossil first described in 1909 by John Smith in a book entitled *Upland Fauna of the Old Red Sandstone Formation of Carrick, Ayrshire*. John Smith wrote a series of books on Ayrshire, ranging from Prehistoric Man to Botany, Drift Deposits and even Conodonts – he was a man of many interests who clearly enjoyed writing. His 'Upland Fauna' consists entirely of trace fossils that he found in small patches of sedimentary rocks that filled hollows and fissures in the Old Red Sandstone lavas of NW Carrick. He produced many names for the traces, but only a few of his names survive to the present day. The fact that he presented his specimens to the Geological Survey has enabled continued research, despite the fact that some are missing. The missing material was on loan at Mainz in Germany, and was destroyed in the Second World War.

The ichnogenus *Siskemia* consists of a central double groove, with a row of imprints on each side. Smith named seven species, but his descriptions are frequently inadequate for modern work, and he does not define specimen numbers in his book. However, some of his material survives in the Geological Survey collections, and from this Walker (1985) was able to redefine the genus.

The double groove in the middle of the trackway represents the drag mark from a double tail spine. When the animal turned a corner the animal lifted its tail, putting it down again in the new direction. From this information it can be inferred that the animal was fairly short, and used about four pairs of legs for walking. *Siskemia* is the trackway of an arthropod, one of many that inhabited the rough surfaces of the Devonian lava fields of Carrick. *Siskemia* has also been found in the Lower Old Red Sandstone at Ferryden, Montrose, and on the island of Kerrera off Oban.

102 *Siskemia* tracks made by a small arthropod dragging a tail with twin spines.

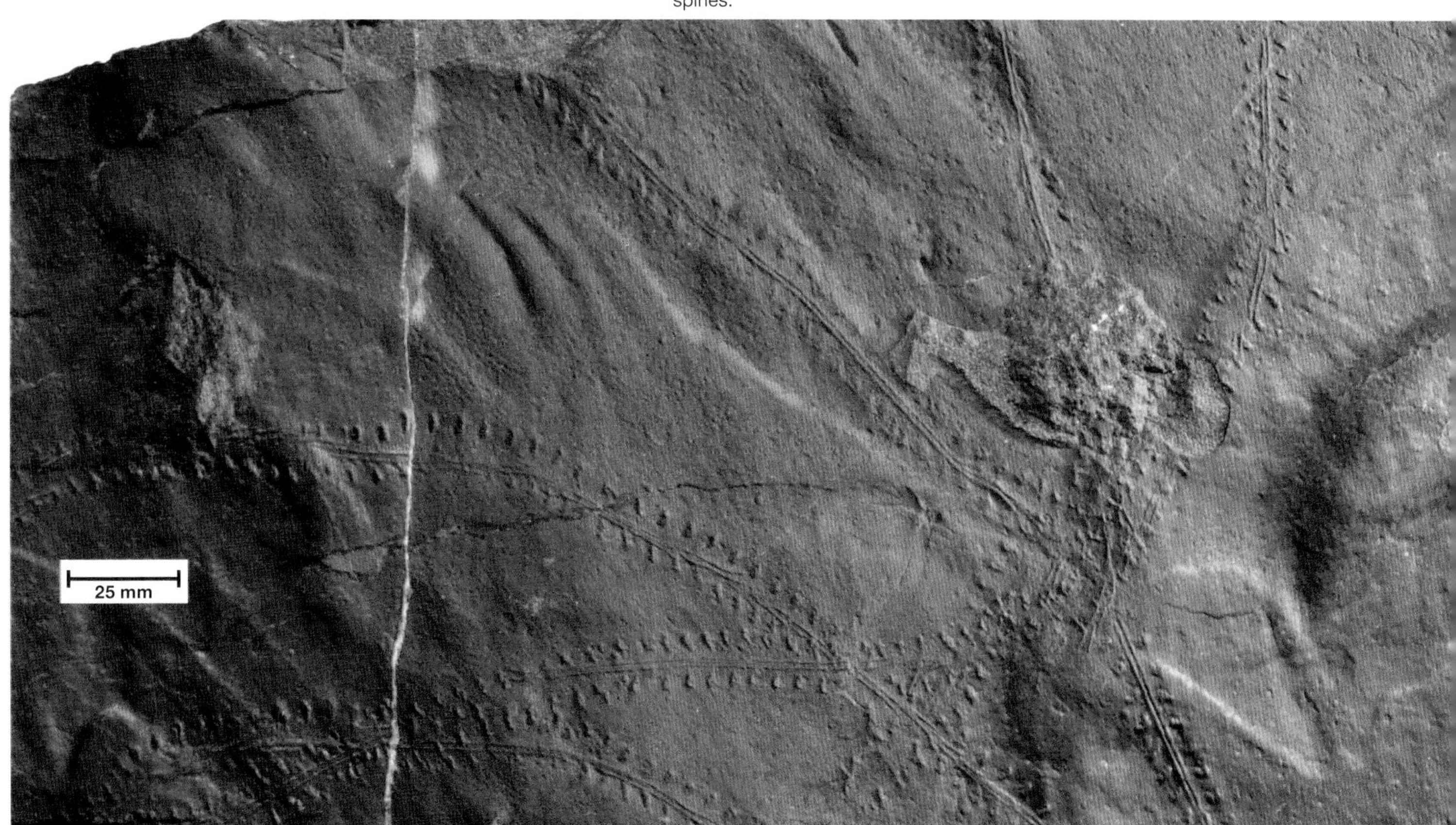

103

Reptile trackway / Trace Fossil / Permian

Trace fossil

Gen et sp	***Chelichnus***
Locality	**Clashach Quarry, Hopeman, Moray**
Age	**Late Permian**
Stratigraphy	**Hopeman Sandstone Formation**

Numerous trackways made by reptiles have been found during quarrying operations at Clashach Quarry. They occur on surfaces of sandstone beds deposited as part of a dune system that fringed the Moray Firth Basin in Permian times. This quarry produced the facing stone to the extension of the Royal Museums Scotland building in Edinburgh, and during quarrying numerous reptile trackways were discovered and documented by Carol Hopkins, as recounted by Clark (2008). The most common trackways are known as *Chelichnus*, and well-preserved examples show the impressions of toes and claws. They were probably made by dicynodont reptiles (see *Gordonia*) Other trackways have been found that show drag marks of a long tail and were made by a reptile similar in form to a monitor lizard. The animals frequently crossed the bare sandy surfaces of the sand dunes, maybe when travelling between areas with food plants and water. The trackways reveal that a range of reptiles inhabited the area, but the only body fossil known is the skull of *Gordonia*.

There is a display of sandstone slabs with trackways near the quarry entrance, and good examples can be seen at Elgin Museum. Similar trackways are also found in the Permian Corncockle Sandstone near Dumfries, and a fine specimen can be seen in the Royal Museum of Scotland in Edinburgh. The Permian trackways from Scotland have been discussed by McKeever (1994) in terms of animal behaviour, but he concludes they are not useful for correlation of Permian strata.

103 *Cheilichnus*, the trackway of a large reptile walking on Permian dune sands at Hopeman.

104

Dinosaur trackways / Trace Fossil / Jurassic

Trace fossil	
Gen et sp	**Dinosaur trackway, similar to *Gigandipus***
Locality	**Staffin, Skye**
Age	**Jurassic, Bathonian**
Stratigraphy	**Duntulm Formation, Great Estuarine Group**

The first dinosaur footprint from the Isle of Skye was described by Andrews and Hudson (1984), and is a large footprint 49cm long, probably made by a herbivorous ornithopod dinosaur with broad toes. The example illustrated shows even larger three-toed prints found on an exposed bedding plane on the shore near Staffin. Fifteen prints were found, with the largest over 50cm long. A few of the prints form parts of trackways. The long, narrow toe imprints with indications of claws indicated to Clark *et al.* (2004) that the maker was probably a large carnivorous therapod dinosaur. The dinosaur was walking on a sandy surface covered by ripple marks produced by waves in shallow water. The sand also contains numerous burrows made by worms and arthropods. The environment was a sandy shore, probably on the edge of a lagoon rather than a seashore. This is the type of environment where trackways can be preserved. The dinosaurs may have hunted and scavenged along the lagoon shore.

Other dinosaur trackways from Skye occur on surfaces with mudcracks, indicating mudflats that had dried out in the sun. It is also possible that the animals had to cross such areas to find water. Dinosaur tracks of various sizes, including some so small they could have been made by small juveniles, have been found at several levels within the Great Estuarine Group on Skye. A wide variety of dinosaurs lived in the area during the Middle Jurassic, but we do not have body fossils (bones) that enable us to identify the makers in more than general terms.

104 Dinosaur trackway on a rippled sand surface in the Jurassic of Skye.

Image Credits

3, 6a,b, 22a,b, Hans Kerp, Münster University.
5b, 67a, Aberdeen University, Rhynie website.
7, Reproduced by courtesy of the British Geological Survey © NERC all rights reserved. IPR/149-03C.
14a,b Brian Bell, Glasgow University.
14c, from Macculloch (1819).
46, 48, 69, 86a, © Hunterian Museum, University of Glasgow 2012, with permission.
63, 76a, 77a, Roger Jones.
71, Andrew Ross, Royal Museums, Scotland.
72a, Courtesy Euan Clarkson, Edinburgh University.
79b, modified after Hemmings (1978).
81b, from Ahlberg and Trewin (1995), with permission of the Royal Society of Edinburgh.
82b, modified from Jarvik (1948).
83a, Stuart Allison, St Andrews University.
86b, Mike Coates, Chicago University.
90b, Mike Coates, Chicago University. with permission of the Royal Society of Edinburgh.
88b from Milner and Sequeira (1994), with permission of the Royal Society of Edinburgh.
89a Jenny Clack, Cambridge University.
89b Dmitry Bogdanov.
91b Carol Hopkins.
92a,b Bill Dalgarno, Elgin Museum.
93a, 94a, Janet Trythall, Elgin Museum.
94b,c, Laura Säilä, in Säilä (2010) with permission of the Royal Society of Edinburgh.
95a, 96a, 102, 104, Neil Clark, Hunterian Museum, University of Glasgow.
97, © Royal Museums, Scotland, with permission.
101, Ted Kellock.

All other images by the author.

Specimen Locations

3, 6a,b, 22a,b,c. Munster University.
5a,b,c, 16a,b, 17c, 28, 32, 39, 47, 64a, 66, 67a,b, 98, 99, University of Aberdeen.
8a, 9, 10a,b, 11, 13, 17a,b, 25, 31, 35, 41, 44, 50, 51, 53, 54, 65, 73a, 74a, 75a, 87, 88a, 90a, 97, Royal Museums, Scotland.
46, 48, 69, 86a, 89a, 91a,b, 95a, 96a, 102, Hunterian Museum, University of Glasgow.
62, Euan Clarkson.
63, 76a, 77a, Roger Jones collection.
71, East Ayrshire Council, Dick Institute, Kilmarnock.
72a, British Geological Survey, Edinburgh.
75b, 78b, Bob Davidson collection.
83a, St Andrews University.
85, Dundee Art Galleries and Museum (Dundee City Council).
92a,b, 93a, 94a,b, Elgin Museum.
All others are in author's collection.

5c, 67b, 75b, are models by Stephen Caine.
78b, model by Paul de Buisonjé.
7, Victoria Park, Glasgow.
14a,b,c, 20c, 43b, 100, 101, 104, specimen *in situ* at locality.
103, open air display near Clashach Quarry, Hopeman.

References

AGASSIZ, L. (1844) *Monographie des Poissons Fossiles du Vieux Grès Rouge ou Système Dévonien (Old Red Sandstone) des Iles Britanniques et de Russie*. xxxvi + 171 p., 43 pls, Neuchâtel.

AHLBERG, P.E. (1991) Tetrapod or near-tetrapod fossils from the Upper Devonian of Scotland. *Nature* **354**, 298–301.

AHLBERG, P.E. (1995) *Elginerpeton pancheni* and the earliest tetrapod clade. *Nature* **373**, 420–25.

AHLBERG, P.E. and TREWIN, N.H. (1995) The postcranial skeleton of the Middle Devonian lungfish *Dipterus valenciennesi*. *Transactions of the Royal Society of Edinburgh: Earth Sciences* **85**, 159–75.

ANDERSON, J. (1859) *Dura Den, A monograph of the yellow sandstone and its remarkable fossil remains*. T. Constable, Edinburgh.

ANDERSON, L.I. and TREWIN, N.H. (2003) An early Devonian arthropod fauna from the Windyfield cherts, Aberdeenshire, Scotland. *Palaeontology* **46**, 467–509.

ANDERSON, L.I., DUNLOP, J.A. and TREWIN, N.H. (2000) A Middle Devonian chasmataspid arthropod from Achanarras Quarry, Caithness, Scotland. *Scottish Journal of Geology* **36**, 151–8.

ANDREWS, J.E. and HUDSON, J.D. (1984) First Jurassic dinosaur footprint from Scotland. *Scottish Journal of Geology* **20**, 129–34.

ANDREWS, S.M. (1982) *The Discovery of Fossil Fishes in Scotland up to 1845, with Checklists of Agassiz's Figured Specimens*. Royal Scottish Museum Studies. Royal Scottish Museum, Edinburgh.

BELL, B.R. and WILLIAMSON, I.T. (2002) Tertiary igneous activity. *In* N.H. Trewin, (ed.) *The Geology of Scotland* (4th edn), 371–407. The Geological Society, London.

BENTON, M.J. (1983) Progressionism in the 1850s: Lyell, Owen, Mantell and the Elgin reptile *Leptopleuron (Telerpeton)*. *Archives of Natural History* **11**, 123–36.

BENTON, M.J. (2009) *Vertebrate Palaeontology (3rd edn)*. Blackwell Science.

BENTON, M.J. and TREWIN, N.H. (1980) *Dictyodora* from the Silurian of Peeblesshire, Scotland. *Palaeontology* **23**, 501–13.

BENTON, M.J. and WALKER, A.D. (1985) Palaeoecology, taphonomy, and dating of Permo-Triassic reptiles from Elgin, north-east Scotland. *Palaeontology* **28**, 207–34.

BENTON, M.J., MARTILL, D.M. and TAYLOR, M.A. (1995) The first Lower Jurassic dinosaur from Scotland: limb bone of a ceratosaur theropod from Skye. *Scottish Journal of Geology* **31**, 177–82.

BLUCK, B.J. (2002) The Midland Valley Terrane. *In* N.H. Trewin, (ed.) *The Geology of Scotland* (4th edn), 149–66. The Geological Society, London.

BOULTER, M.C. and KVAČEK, Z. (1989) The Palaeocene flora of the Isle of Mull. *Special Papers in Palaeontology* **42**. Palaeontological Association.

BRIGGS, D.E.G. and CLARKSON, E.N.K. (1985) The Lower Carboniferous shrimp *Tealliocaris* from Gullane, East Lothian. *Transactions of the Royal Society of Edinburgh: Earth Sciences* **76**, 173–201.

BRIGGS, D.E.G., CLARKSON, E.N.K. and ALDRIDGE, R.A. (1983) The conodont animal. *Lethaia* **16**, 1–14.

BRIDGLAND, D.R., SAVILLE, A. and SINCLAIR, J.M. (1997) New evidence for the origin of the Buchan Ridge Gravel, Aberdeenshire. *Scottish Journal of Geology* **33**, 43–50.

CATER, J.M.L., BRIGGS, D.E.G. and CLARKSON, E.N.K. (1989) Shrimp-bearing sedimentary successions in the Lower Carboniferous (Dinantian) Cementstone and oil Shale Groups of northern Britain. *Transactions of the Royal Society of Edinburgh: Earth Sciences* **80**, 5–15.

CHEN, P-J. and HUDSON, J.D. (1991) The conchostracan fauna of the Great Estuarine Group, Middle Jurassic, Scotland. *Palaeontology* **34**, 515–45.

CLACK, J.A. (2002) An early tetrapod from 'Romer's Gap'. *Nature* **418**, 72–6.

CLACK, J.A. and FINNEY, S.M. (2005) *Pederpes finneyae*, an articulated tetrapod from the Tournaisian of western Scotland. *Journal of Systematic Palaaeontology* **2**, 311–46.

CLARK, N.D.L. (1991) *Palaemysis dunlopi* Peach 1908 (Eocarida, Crustacea) from the Namurian (Carboniferous) of the western Midland Valley. *Scottish Journal of Geology* **27**, 1–10.

CLARK, N.D.L. (2008) The Elgin Marvels. *Deposits* **13**, 36–9.

CLARK, N.D.L. and BARCO RODROGUEZ, J.L. (1998) The first dinosaur trackway from the Valtos Sandstone Formation (Bathonian, Jurassic) of the Isle of Skye, Scotland, UK. *Geogaceta* **24**, 79–82.

CLARK, N.D.L., BOOTH, P., BOOTH, C. and ROSS, D.A. (2004) Dinosaur footprints from the Duntulm Formation (Bathonian, Jurassic) of the Isle of Skye. *Scottish Journal of Geology* **40**, 13–21.

CLARK, N.D.L., BOYD, J.D., DIXON, R.J. and ROSS, D.A. (1995) The first Middle Jurassic dinosaur from Scotland: a cetiosaurid? (Sauropoda) from the Bathonian of the Isle of Skye. *Scottish Journal of Geology* **31**, 171–6.

CLARKSON, E.N.K. (1966) Schizochroal eyes and vision of some Silurian acastid trilobites. *Palaeontology* **9**, 1–29.

CLARKSON, E.N.K. (1985) A brief history of Scottish Palaeontology 1834–1984. *Scottish Journal of Geology* **21**, 389–406.

CLARKSON, E.N.K. (1998) *Invertebrate Palaeontology and Evolution*. (4th edn). Chapman and Hall, London.

CLARKSON, E.N.K. (2000) Pentland Odyssey. *Scottish Journal of Geology* **36**, 8–16.

CLARKSON, E.N.K. and TAYLOR, C.M. (1992) Dob's Linn. *In* A.D.

McAdam, E.N.K. Clarkson and P. Stone (1992) *Scottish Borders Geology: an excursion guide.* Scottish Academic Press, Edinburgh.
CLARKSON, E.N.K. and TRIPP, R.P. (1982) The Ordovician trilobite *Calyptaulax brongniartii* (Portlock) *Transactions of the Royal Society of Edinburgh: Earth Sciences* **72,** 287–94.
CLARKSON, E.N.K., HARPER, D.A.T., TAYLOR, C.M. and ANDERSON, L.I. (2008) *Silurian Fossils of the Pentland Hills, Scotland.* Field Guides to Fossils 11, The Palaeontological Association.
CLARKSON, E.N.K., MILNER, A.R. and COATES, M.I. (1994) Palaeoecology of the Viséan of East Kirkton, West Lothian, Scotland. *Transactions of the Royal Society of Edinburgh: Earth Sciences* **84,** 417–25.
CLEAL, C.J. and THOMAS, B.A. (1994) *Plant fossils of the British Coal Measures.* Palaeontological Association Field Guides to Fossils, Number 6. Palaeontological Association, London.
CLEAL, C.J. and THOMAS, B.A. (1999) *Plant fossils: the history of land vegetation.* The Boydell Press, Woodbridge.
COATES, M.I. and SEQUEIRA, S.E.K. (2001) A new stethacanthid chondrichthyan from the Lower Carboniferous of Bearsden, Scotland. *Journal of Vertebrate Palaeontology* **21**, 439–59.
COWIE, J. and McNAMARA, K.J. (1978) *Olenellus* (Trilobita) from the Lower Cambrian strata of north-west Scotland. *Palaeontology* **21**, 615–34.
DAVIDSON, R.G. and NEWMAN, M.J. (2003) James Powrie, chronicler of the Scottish Lower Devonian. *Proceedings of the Geologists Association* **114**, 243–6.
DAWSON, W. and PENHALLOW, D.P. (1891) *Parka decipiens,* notes on specimens from the collections of James Reid Esq. *Transactions of the Royal Society of Canada* **9**, 3, Pl 1.
DENISON, R. (1979) Volume 5 Acanthodii. *In* H-P. Schultze (ed.) *Handbook of Paleoichthyology*. Gustav Fisher Verlag.
DON, A.W.R. and HICKLING, G. (1917) On *Parka decipiens. Quarterly Journal of the Geological Society of London* **71**, 648–66.
DUFF, P. (1842) *Sketch of the Geology of Moray*. Elgin.
EDWARDS, D. (2004) Embryophytic sporophytes in the Rhynie and Windyfield cherts. *Transactions of the Royal Society of Edinburgh: Earth Sciences* **94,** 397–410.
EDWARDS, D.S. (1986) *Aglaophyton major,* a non-vascular land plant from the Devonian Rhynie Chert. *Botanical Journal of the Linnean Society* **93**, 173–204.
FALCON-LANG, H.J. (2008) Marie Stopes and the Jurassic floras of Brora, NE Scotland. *Scottish Journal of Geology* **44**, 65–73.
FAYERS, S.R. and TREWIN, N.H. (2004) A review of the palaeoenvironments and biota of the Windyfield chert. *Transactions of the Royal Society of Edinburgh: Earth Sciences* **94,** 325–39.
FAYERS, S.R., DUNLOP, J.A. and TREWIN, N.H. (2005) A new early Devonian Trigonotarbid Arachnid from the Windyfield chert, Rhynie, Scotland. *Journal of Systematic Palaeontology* **2**, 269–84.
FERGUSON, W. (1850) Notice of the occurrence of Chalk flints and Greensand fossils in Aberdeenshire. *Philosophical Magazine* **37**, 430–8.
FLEMING, J. (1830) On the occurrence of vertebrated animals in the Old Red Sandstone of Fifeshire. *Cheek's Edinburgh Journal of Natural and Geographical Science* **3**, 81–6.
FORSYTH, I.H, and CHISHOLM, J.I. (1977) *The Geology of East Fife.* Memoir of the Geological Survey of Great Britain.
GEIKIE, A. (1902) *The Geology of East Fife.* Memoir of the Geological Survey of Great Britain.
HALL, A.M. (1993) Moss of Cruden. *In* J.E. Gordon and D.G. Sutherland (eds) *Quaternary of Scotland.* Geological Conservation Review Series **6**, 218–21.
HARPER, D.A.T. (1982a) The late Ordovician Lady Burn Starfish Beds of Girvan. *Proceedings of the Geological Society of Glasgow* **122/123**, 28–32.
HARPER, D.A.T. (1982b) The stratigraphy of the Drummock Group (Ashgill), Girvan. *Geological Journal* **17**, 251–277.
HEMMINGS, S.K. (1978) The Old Red Sandstone antiarchs of Scotland: *Pteriychthyodes* and *Microbrachius. Palaeontographical Society Monograph* **131**, 1–64.
HEMSLEY, A.R. (1990) *Parka decipiens* and land plant spore evolution. *Historical Biology* **4**, 39–50.
HESSELBO, S.P. and TREWIN, N.H. (1984) Deposition, diagenesis and structures of the Cheese Bay Shrimp Bed, Lower Carboniferous, East Lothian. *Scottish Journal of Geology* **20**, 281–96.
HESSELBO, S.P., OATES, M.J. and JENKYNS, H.C. (1998) The Lower Lias Group of the Hebrides Basin. *Scottish Journal of Geology* **34**, 23–60.
HUDSON, J.D. and TREWIN, N.H. (2002) Jurassic. *In* N.H. Trewin, (ed.) *The Geology of Scotland* (4th edn), 323–50. The Geological Society, London.
HUXLEY, T.H. (1859) On the *Stagonolepis robertsoni* (Agassiz) of the Elgin Sandstones: and on the recently discovered footmarks in the sandstones of Cummingstone. *Quarterly Journal of the Geological Society, London* **15**, 440–60.
HUXLEY, T.H. and SALTER, J.W. (1859) *British Organic Remains, Monograph 1.* Memoir of the Geological Survey, U.K.
JARVIK, E. (1948) On the morphology and taxonomy of the Middle Devonian Osteolepid fishes of Scotland. *Kungliga Svenska Vetenskapsakademiens Handlingar,* **3**, (25), 1, 1–301.
JEFFERIES, R.P.S (1986) *The Ancestry of the Vertebrates*. British Museum (Natural History), London.
JERAM, A.J. (1994) Scorpions from the Viséan of East Kirkton, West Lothian, Scotland, with a revision of the infraorder Mesoscorpionina. *Transactions of the Royal Society of Edinburgh: Earth Sciences* **84,** 283–99.
JUDD, J.W. (1873) The secondary rocks of Scotland. *Quarterly Journal of the Geological Society, London* **29**, 97–195.
KAMMER, T.W. and AUSICH, W.I. (2007) Soft tissue preservation of the hind gut in a new genus of cladid crinoid from the Mississippian (Viséan, Asbian) at St Andrews, Fife. *Palaeontology* **50**, 951–9.
KELMAN, R., FEIST, M., TREWIN, N.H. and HASS, H. (2004)

Charophyte algae from the Rhynie chert. *Transactions of the Royal Society of Edinburgh: Earth Sciences* **94**, 445–55.

KIDSTON, R. and LANG, W.H. (1917–1921) On Old Red Sandstone plants showing structure from the Rhynie chert bed, Aberdeenshire (Parts 1–5). *Transactions of the Royal Society of Edinburgh* **51**, 643–80; **52**, 603–27, 643–80, 831–54, 855–902.

LANKESTER, E.R. (1868, 1870) *A monograph of the fishes of the Old Red Sandstone of Britain. Part 1: The Cephalaspidae* Palaeontographical Society (Monographs) Sections 1 (1868), 2 (1870).

LAPWORTH, C. (1878) The Moffat Series. *Quarterly Journal of the Geological Society, London* **34**, 240–346.

LAPWORTH, C. (1882) The Girvan succession. Part 1. Stratigraphy. *Quarterly Journal of the Geological Society*, London **38**, 537–664.

LAPWORTH, C. (1888) On the discovery of the *Olenellus* fauna in the Lower Cambrian rocks of Britain. *Geological Magazine* **5**, 484.

LEE, G.W. (1925) Mesozoic rocks of East Sutherland and Ross. *In* H.H. Read, G. Ross and J. Phemister. *The geology of the country around Golspie, Sutherland*. Memoirs of the Geological Survey 65–115.

LISTON, J.J. (2004) A re-examination of a Middle Jurassic sauropod limb bone from the Bathonian of the Isle of Skye. *Scottish Journal of Geology* **40**, 119–22.

LYELL, C. (1841) *Elements of Geology* (2nd edn).

MACCULLOCH, J. (1819) *A Description of the Western Islands of Scotland, Including the Isle of Man, Comprising an account of their Geological Structure; With Remarks on their Agriculture, Scenery and Antiquities.* 3 Vols. Constable, London.

MACDONALD, A.C. and TREWIN, N.H. (2009) The Upper Jurassic of the Helmsdale area. *In* N.H. Trewin and A. Hurst (eds) *Excursion Guide to the geology of East Sutherland and Caithness*. Dunedin Academic Press, Edinburgh.

MACGREGOR, A.R. (1973) *Fife and Angus Geology* (2nd edn) Scottish Academic Press, Edinburgh.

McADAM, A.D., CLARKSON, E.N.K. and STONE, P. (1992) *Scottish Borders Geology: an excursion guide*. Scottish Academic Press, Edinburgh.

McKEEVER, P.J. (1994) The behavioral and biostratigraphical significance and origin of vertebrate trackways from the Permian of Scotland. *Palaios* **9**, 477–87.

MACKIE, W. (1913) The rock series of Craigbeg and Ord Hill, Rhynie, Aberdeenshire. *Transactions of the Edinburgh Geological Society* **10**, 205–36.

MANTELL, G.A. (1852) Description of *Telerpeton elginense*, a fossil reptile recently discovered in the Old Red Sandstone of Moray: with observations on supposed fossil ova of batrachians in the Lower Devonian strata of Forfarshire. *Quarterly Journal of the Geological Society, London* **8**, 100–9.

MILES, R.S. and WESTOLL, T.S. (1963) Two new genera of Coccosteid Arthrodira from the Middle Old Red Sandstone of Scotland and their stratigraphic distribution. *Transactions of the Royal Society of Edinburgh* **65**, 179–210.

MILES, R.S. and WESTOLL, T.S. (1968) The placoderm fish *Coccosteus cuspidatus* Miller ex Agassiz from the Middle Old Red Sandstone of Scotland. Pt 1 Descriptive morphology. *Transactions of the Royal Society of Edinburgh* **67**, 373–476.

MILLER, H. (1841) *The Old Red Sandstone*. John Johnstone, Edinburgh.

MILLER, H. (1842) *The Old Red Sandstone.* (2nd edn) John Johnstone, Edinburgh.

MILLER, H. (1849.)*Footprints of the Creator, or the Asterolepis of Stromness.* Johnstone and Hunter, London and Edinburgh.

MILLER, H. (1857) *The testimony of the rocks*. Thomas Constable & Co., Edinburgh.

MILLER, H. (1858) *The Cruise of the Betsey, with Rambles of a Geologist.* Constable, Edinburgh.

MILLER, H. (1859) *Sketch book of popular geology*. Thomas Constable and Company, Edinburgh

MILNER, A.R. and SEQUEIRA, S.E.K. (1994) The temnospondyl amphibians from the Viséan of East Kirkton, West Lothian, Scotland. *Transactions of the Royal Society of Edinburgh: Earth Sciences* **84**, 331–61.

MORTON, N. (1965) The Bearreraig Sandstone Series (Middle Jurassic) of Skye and Raasay. *Scottish Journal of Geology* **1**, 189–216.

MURCHISON, R.I. (1839) *The Silurian System*. Murray, London.

MURCHISON, R.I. (1867) *Siluria* (4th edn). Murray, London.

NEWMAN, M.J. and DEN BLAAUWEN, J.L. (2008) New information on the enigmatic Devonian vertebrate *Palaeospondylus gunni. Scottish Journal of Geology* **44**, 89–91.

NEWMAN, M.J. and TREWIN, N.H. (2001) A new jawless vertebrate from the Middle Devonian of Scotland. *Palaeontology* **44**, 43–51.

NEWMAN, M.J. and TREWIN, N.H. (2008) Discovery of the arthrodire genus *Actinolepis* (class Placodermi) in the Middle Devonian of Scotland. *Scottish Journal of Geology* **44**, 83–8.

NICHOLSON, H.A. and ETHERIDGE, R. (Jun) (1878–80) *A monograph of the Silurian fossils of the Girvan district in Ayrshire, with reference to those contained in the 'Gray Collection'*. Blackwood and Sons, Edinburgh and London 1–341.

NICOL, J. (1841) On the geology of Peeblesshire. *Transactions of the Highland and Agricultural Society of Scotland* **14**, 149–206.

NICOL, J. (1844) *Guide to the Geology of Scotland*. Oliver and Boyd, Edinburgh.

NIKLAS, K.J. (1976) Morphological and ontogenetic reconstructions of *Parka decipiens* Fleming and *Pachytheca* Hooker from the Lower Old Red Sandstone, Scotland. *Transactions of the Royal Society, Edinburgh* **69**, 483–98.

OLIVER, G.J.H., STONE, P. and BLUCK, B.J. (2002) *In* N.H. Trewin (ed.) *The Geology of Scotland* (4th edn), 167–200. The Geological Society, London.

OWEN, R. (1851) Vertebrate air-breathing life in the Old Red Sandstone.

Literary Gazette, and Journal of Belles Lettres, December 20, 1851, 2.

PEACH, B.N. (1894) Additions to the fauna of the Olenellus zone of the North-West Highlands. *Quarterly Journal of the Geological Society, London* **50**, 661.

PEACH, B.N. and HORNE, J. (1899) *The Silurian rocks of Britain. Vol 1 Scotland.* Memoirs of the Geological Survey of the United Kingdom.

PEACH, B.N., HORNE, J., GUNN, W., CLOUGH, C.T. and HINXMAN, L.W. (1907) *The Geological Structure of the North-West Highlands of Scotland.* Memoirs of the Geological Survey of Great Britain.

POINAR, Jr. G., KERP, H. and HASS, H. (2008) *Palaeonema phyticum gen. n., sp. n.* (Nematoda: Palaeonematidae *fam. n.*), a Devonian nematode associated with early land plants. *Nematology* **10**, 9–14.

POWRIE, J. (1864) On the fossiliferous rocks of Forfarshire and their contents. *Quarterly Journal of the Geological Society of London* **20**, 413–29.

RAMSBOTTOM, W.H.C. (1961) *A monograph of the British Ordovician Crinoidea.* Monograph of the Palaeontographical Society **114**, 1–36.

RITCHIE, A. (1968) New evidence on *Jamoytius kerwoodi* White, an important ostracoderm from the Silurian of Lanarkshire, Scotland. *Palaeontology* **11**, 21–39.

ROGERS, D.A. (1990) Probable tetrapod tracks rediscovered in the Devonian of N Scotland. *Journal of the Geological Society, London* **147**, 746–8.

ROLFE, W.D.I., CLARKSON, E.N.K. and PANCHEN, A.L. (eds) (1994) Volcanism and early terrestrial biotas. *Transactions of the Royal Society of Edinburgh: Earth Sciences* **84,** (3,4). Proceedings of conference, 1992, Edinburgh.

ROSS, A. (2010) A review of the Carboniferous insect fossils from Scotland. *Scottish Journal of Geology* **46**, 157–68.

SÄILÄ, L.K. (2010) Osteology of *Leptopleuron lacertinum* Owen, a procolophonoid parareptile from the Upper Triassic of Scotland, with remarks on ontogeny, ecology and affinities. *Earth and Environmental Science Transactions of the Royal Society of Edinburgh* **101**, 1–25.

SANSOM, R.S. (2009) Phylogeny, classification and character polarity of the Osteostraci (Vertebrata) *Journal of Systematic Palaeontology* **7**, 95–115.

SAXON, J. (1975) *The fossil fishes of the North of Scotland* (2nd edn). 1–49. Caithness Books, Thurso.

SMITH, J. (1909) Upland fauna of the Old Red Sandstone of Carrick, Ayrshire. A.W. Cross, Kilwinning. 1–41, pls 1–9.

SMITH, W. (1816) *Strata identified by organised fossils.* London.

SMITHSON, T.R. and ROLFE, W.D.I. (1990) *Westlothiana* gen. nov.: naming the earliest known reptile. *Scottish Journal of Geology* **26**, 157–38.

SMITHSON, T.R., CARROLL, R.L., PANCHEN, A.L. and ANDREWS, S.M. (1994) *Westlothiana lizziae* from the Viséan of East Kirkton, West Lothian, Scotland and the amniote stem. *Transactions of the Royal Society of Edinburgh: Earth Sciences,* **84,** 383–412.

STENSIO, E.A. (1932) *The cephalaspids of Great Britain.* Monograph of the British Museum (Natural History), London 1–220, pls1–65.

STONE, P., RIGBY, S. and FLOYD, J.D. (2003) Advances in graptolite biostratigraphy: an introduction. *Scottish Journal of Geology* **39**, 11–15.

STOPES, M.C. (1907) The flora of the Inferior Oolite of Brora. *Quarterly Journal of the Geological Society of London* **63**, 375–82.

STOPES, M.C. (1910) *Ancient Plants.* Blackie and Son Ltd., London.

TAYLOR, M.A. (2007) *Hugh Miller, stonemason, geologist, writer.* National Museums Scotland.

THOMSON, C.A. and WILKINSON, I.P. (2009) Robert Kidston (1852–1924): biography of a Scottish palaeobotanist. *Scottish Journal of Geology* **45**, 161–8.

THOMSON, K.S., SUTTON, M. and THOMAS, B. (2003) A larval Devonian lungfish. *Nature* **426**, 833–4.

TRAQUAIR, R.H. (1890) On the fossil fishes of Achanarras Quarry, Caithness. *Annals and Magazine of Natural History, series 6.* **6**, 479–86.

TRAQUAIR, R.H. (1894–1914) *A monograph of the fishes of the Old Red Sandstone of Britain. Part 2: The Asterolepidae.* Palaeontographical Society (Monographs) Sections 1–4.

TREWIN, N.H. (1986) Palaeoecology and sedimentology of the Achanarras fish bed of the Middle Old Red Sandstone, Scotland. *Transactions of the Royal Society of Edinburgh: Earth Sciences* **77**, 21–46.

TREWIN, N.H. (1994) Depositional environment and preservation of biota in the Lower Devonian hot-springs of Rhynie, Aberdeenshire, Scotland. *Transactions of the Royal Society of Edinburgh: Earth Sciences* **84**, 433–42.

TREWIN, N.H. (ed.) (2002) *The Geology of Scotland.* (4th edn) Geological Society, London.

TREWIN, N.H. (2004) History of research on the geology and palaeontology of the Rhynie area, Aberdeenshire, Scotland. *Transactions of the Royal Society of Edinburgh: Earth Sciences* **94**, 285–97.

TREWIN, N.H. (2008) *Fossils Alive! or new walks in an old field.* Dunedin Academic Press, Edinburgh.

TREWIN, N.H. (2009) The Old Red Sandstone of Caithness. *In* N.H. Trewin and A. Hurst (eds) *Excursion Guide to the geology of East Sutherland and Caithness.* Dunedin Academic Press, Edinburgh.

TREWIN, N.H. and DAVIDSON, R.G. (1996) An Early Devonian lake and its associated biota in the Midland Valley of Scotland. *Transactions of the Royal Society of Edinburgh: Earth Sciences* **86**, 233–46.

TREWIN, N.H. and KNELLER, B.C. (1987a) Old Red Sandstone and Dalradian of Gamrie Bay. *In* N.H. Trewin, B.C. Kneller and C. Gillen (eds), *Excursion guide to the geology of the Aberdeen area.* 113–126. Scottish Academic Press.

TREWIN, N.H. and KNELLER, B.C. (1987b) The Lower Old Red Sandstone of New Aberdour. *In* N.H. Trewin, B.C. Kneller and C. Gillen, (eds). *Excursion Guide to the geology of the Aberdeen area,* Scottish Academic press, Edinburgh, 131–42.

TREWIN, N.H. and RICE, C.M. (eds) (2004) The Rhynie hot-spring system: geology, biota and mineralisation. Proceedings of a conference

at Aberdeen, September 2003. *Transactions of the Royal Society of Edinburgh: Earth Sciences* **94**, (4).
TREWIN, N.H. and ROLLIN, K.E. (2002) Geological history and structure of Scotland. *In* N.H. Trewin (ed.) *The Geology of Scotland* (4th edn), 1–26. The Geological Society, London.
TREWIN, N.H. and THIRLWALL, M.F. (2002) The Old Red Sandstone. *In* N.H. Trewin (ed.) *The Geology of Scotland* (4th edn), 213–50. The Geological Society, London.
TREWIN, N.H. and HURST, A. (eds) (2009) *Excursion Guide to the geology of East Sutherland and Caithness*. Dunedin Academic Press, Edinburgh.
TREWIN, N.H., FAYERS, S.R. and ANDERSON, L.I. (2001) The Biota of Early Terrestrial Ecosystems: The Rhynie Chert. www.abdn.ac.uk/rhynie
VAN DER BURGH, J. and VAN KONIJNENBURG-VAN CITTERT, J.H.A. (1984) A drifted flora from the Kimmeridgian (Upper Jurassic) of Lothbeg Point, Sutherland, Scotland. *Reviews of Palaeobotany and Palynology* **43**, 359–96.
WALKER, E.F. (1985) Arthropod ichnofauna of the Old Red Sandstone at Dunure and Montrose, Scotland. *Transactions of the Royal Society of Edinburgh: Earth Sciences* **76**, 287–97.
WILSON, H.M. and ANDERSON, L.I. (2004) Morphology and taxonomy of Palaeozoic millipedes (Diplopoda: Chilognatha: Archipolypoda) from Scotland. *Journal of Palaeontology* **78**, 169–184.
WOOD, S.P. (1982) New basal Namurian (Upper Carboniferous) fishes and crustaceans found near Glasgow. *Nature* **297**, 574–7.
WRIGHT, J. (1934) New Scottish and Irish fossil crinoids. *Geological Magazine* **71**, 241–268.
WRIGHT, J. (1939) The Scottish Carboniferous Crinoidea. *Transactions of the Royal Society of Edinburgh* **60**, 1–78.
WRIGHT, (1952–60) *The British Carboniferous Crinoidea*. Monograph of the Palaeontographical Society. Parts1,2, 1–347.
WYSE JACKSON, P.N. (2010) *Introducing Palaeontology: A guide to ancient life*. Dunedin Academic Press, Edinburgh.

Biological Index: Fossil and modern organisms

Note: illustrations in **bold**, main sections *italic.*

Gazetteer